Advanced Studies Mobile Research Center Bremen

Series editors
A. Förster, Bremen, Germany
O. Herzog, Bremen, Germany
M. Lawo, Bremen, Germany
H. Witt, Bremen, Germany

Das Mobile Research Center Bremen (MRC) wurde 2004 als Forschungs- und Transferinstitut des Landes Bremen gegründet, um mit dem Technologie-Zentrum Informatik und Informationstechnik (TZI) der Universität Bremen und in enger Zusammenarbeit mit der Wirtschaft mobile Informatik-, Informations- und Kommunikationstechnologien zu erforschen, entwickeln und erproben. Hier wird eine Vielzahl interdisziplinärer Arbeitsgruppen, die sich mit der Entwicklung und Anwendung mobiler Lösungen beschäftigen, vernetzt. Die Reihe „Advanced Studies" präsentiert ausgewählte hervorragende Arbeitsergebnisse aus der Forschungstätigkeit der Mitglieder des MRC.

The Mobile Research Center Bremen (MRC), established 2004 as research cluster of the state of Bremen, investigates, develops and tests mobile computing, information and communication technologies with the Center for Computing and Communication Technologies (TZI) of the University of Bremen in close collaboration with the industry. It links and coordinates interdisciplinary research teams from different universities and institutions, which are concerned with the development and application of mobile solutions. The series "Advanced Studies" presents a selection of outstanding results of MRC's research projects.

More information about this series at http://www.springer.com/series/12158

Michael Lawo · Peter Knackfuß
Editors

Clinical Rehabilitation Experience Utilizing Serious Games

Rehabilitation Technology
and a Technical Concept
for Health Data Collection

Springer Vieweg

Editors
Michael Lawo
Bremen, Germany

Peter Knackfuß
Bremen, Germany

The projects referenced have received funding from the European Union's research and innovation programs under grant agreements FP7-ICT No. 216461 (CHRONIUS), FP7-Health No306113 (Rehab@Home), FP7-ICT No. 287677 (REMPARK), AAL No. 2011-4-094 (SafeMove), FP7-ICT No. 288532 (USEFIL), FP7-ICT No. 248294 (VICON), FP6-IST No. 004216 (wearIT@work) and its related subproject uWear.

Advanced Studies Mobile Research Center Bremen
ISBN 978-3-658-21956-7   ISBN 978-3-658-21957-4   (eBook)
https://doi.org/10.1007/978-3-658-21957-4

Library of Congress Control Number: 2018942027

Springer Vieweg

Printed on acid-free paper

This Springer Vieweg imprint is published by the registered company Springer Fachmedien Wiesbaden GmbH part of Springer Nature
The registered company address is: Abraham-Lincoln-Str. 46, 65189 Wiesbaden, Germany

Preface

This book emerged out of research done during the period from 2004 to 2016 on the topic of mobile and wearable computing. This research only became possible by national and European public funding. Many concepts outlined in this research are today part of daily living such as Smartphones and Wearables.

The beginning of the research focused mainly on identifying the benefits of mobile computing supporting people working on the move. This was where the editors of this book met as managers in 2004 when working in the European funded project wearIT@work which was coordinated by Prof. Dr. Otthein Herzog who brought the enthusiasm for Wearable Computing to us from research results of MIT and CMU during the 1990s. This research initiated the Mobile Research Centre in Bremen and was the start of a series of research projects that evolved from Wearable Computing supporting our daily work towards healthcare solutions for people with chronic disease and/or undergoing rehabilitation.

At the beginning, there seemed to be a need to make a difference between ubiquitous, pervasive, mobile and wearable computing. Today we know that, as a result of the mobile internet, this has all merged and the key issues are solved by platforms and their operating systems supporting all the components alike.

Our research did not solely focus on technical solutions and the search for a general approach but also how people can live with this technology. Thus, social and organisational aspects were part of our research. We made contact with researchers in this field and worked with them continuously throughout our research. However, when moving towards medical support, medical and psychological aspects also needed our attention. We learnt about the opportunities of Serious Games and the need for clinical studies when targeting solutions designed to become part of any kind of therapy.

The problems we intended to solve involved people belonging to a target group that none of the research teams was part of. We needed experience in aircraft maintenance, car assembly, and firefighting. We had to learn about the ward round in a hospital and the lifestyles of people with chronic diseases or those recovering from a stroke. Cultural and emotional issues such as mental disorders we also had to learn about.

All this brought us together with a wonderful team of researchers in Europe and worldwide. It was a very rich time of learning for us and we are grateful that those researchers shared their experiences with us; only through interdisciplinary collaboration could we gain all these experiences.

In the context of the Rehab@Home project, we decided to ask some of our co-workers to contribute to the book that you now hold in your hands. The result is a comprehensive presentation of research results covering different important aspects of research in the domain of Wearable and Pervasive computing aimed at delivering a better life.

The concept of the book is that each book chapter covers one specific aspect. It offers the interested reader references to the literature and/or other chapters of the book. In this way we intend to motivate further study and research.

We thank all our numerous co-workers for their contributions and hope that this book finds interested readers.

Bremen, 2017

Peter Knackfuß and Michael Lawo

Contents

I   Wearable and Pervasive Computing for Healthcare and Towards Serious Games
An Introduction

Michael Lawo / Peter Knackfuß

Abstract

Physical activity is a major part of the user's context for wearable computing applications. When using body-worn sensors, one can acquire the user's physical activities. In recent years, wearable and pervasive computing were the basis of many research projects in industrial and healthcare environments. As such, projects used wearable computing technology to help people with chronic diseases like chronic obstructive pulmonary disease (COPD), chronic kidney disease (CKD), and Parkinson disease (PD). However, the needs of the target groups cannot be satisfied just by technical solutions alone. Motivation plays a major role in really helping people. Serious Games or exergames promise a positive effect on motivation and activity. In this chapter, we will explain the basic technology concepts of wearable and pervasive computing and motivation using Serious Games based on the personal experience of the authors and their co-workers in topic-related research projects over the last twelve years.

© Springer Fachmedien Wiesbaden GmbH, part of Springer Nature 2018
M. Lawo und P. Knackfuß (Hrsg.), *Clinical Rehabilitation Experience Utilizing Serious Games*, Advanced Studies Mobile Research Center Bremen, https://doi.org/10.1007/978-3-658-21957-4_1

1   Motivation

In recent years, wearable and pervasive computing has drastically changed our understanding of technology-based support in our daily life. Sometimes we even have the feeling that it has become too much. Whether in the office, on the move or on the shop floor or hospital; technology provides us with information at any time and at any place.

In this book, we like to draw the attention towards a very specific application domain of wearable and pervasive computing: Healthcare. In healthcare, this kind of technology can help patients as well as caregivers and physicians to master the challenges of the demographic change.

In 1997, the number of people over 65 constituted 6.6% of the world's population, predicted to increase to 10% by 2025. It is likely that this will lead to a rise in demand for long-term residential care. Common elderly diseases may include one or more of the following: arthritis, cancer, cardiovascular (e.g. blood pressure and heart disease), cerebra-vascular (e.g. strokes), dementia, depression, diabetes, falls and injuries, gastrointestinal disorders, hearing impairment, memory, nutrition, osteoporosis, Parkinson's and Alzheimer's diseases, respiratory disease, pressure ulcers, sleep problems, thyroid disease, urinary disorders and visual impairment. In many cases, people experience considerable health gain, both from the physical and cognitive perspectives by successful rehabilitation, which focuses on lessening the impact of specific disabling conditions. Under these circumstances, wearable and pervasive computing as an ambient technology seems an appropriate measure.

In recent years, the authors of this book participated with their co-workers in quite a number of European and national funded research projects based on wearable and pervasive computing and its applications in industrial and healthcare environments.[1] Some of these projects used wearable computing technology to help people with chronic diseases like COPD, CKD, and Parkinson.[2] Also, age-related mild to moderate impairments of sight, hearing and motor abilities and, lately, even mild cognitive impairments were addressed.[3]

---

[1] www.wearitatwork.com, www.siwear.de, www.xpick.de; all accessed 1.3.2016
[2] www.chronious.eu, www.usefil.eu, www.help-project-parkinson.com, www.rempark.eu; all accessed 1.3.2016
[3] www.vicon-project.eu, http://assam.nmshost.de/, www.safemove-project.eu; all accessed 1.3.2016

With all these projects, we got an understanding of the needs of the target group but also the feeling that a technical solution alone was not enough for really helping people. Here the idea of using Serious Games [1] or exergames originated. Literature [2] promises a positive effect from exergames on motivation and activity in general. We therefore integrated this insight into our recent research projects addressing people with slight cognitive impairments and those in rehabilitation after a stroke or having MS.[4] In the following chapter, we will first explain the basic concept behind this technological development and motivation through the use of Serious Games or exergames. We will further summarize the findings of our different projects as the basis of the research outcomes described in the remaining book chapters.

2 Wearable and Pervasive Computing

Mobile and pervasive computing have the common ability to access information at any time. Pervasive computing integrates with the environment; it lets the user access information only in especially prepared environments. Mobile computing goes with the user and does not require any more than just a communication infrastructure and sufficient energy supply. To provide further integration, the system needs to evaluate the user's context of activity e.g. when working, to give appropriate unobtrusive support. With the knowledge of such context, one can integrate this support more seamlessly into different activities than a smartphone. With the capability to integrate electronic components as computers into clothing, they are highly interesting for industrial application domains such as maintenance or assembly. There, workers have to carry out complex activities on real world objects while simultaneously often having to consult instructional material.

Computers supporting workers during such physical activities are called Wearable Computers. They offer portability during operation, enable hands-free or hands-limited use, can sense the user or environment context and run continuously. To let users take advantage of wearable computing technology during their primary activities in the field, one needs a thorough design of the human-computer interface, reflecting its new and unique context of usage in a mobile dual-task situation.

---

[4] www.safemove-project.eu, www.rehabathome-project.eu; all accessed 1.3.2016

Minimized cognitive load and the ability to let users divide their attention between both types of task as easily and efficiently as possible are crucial design issues of user interfaces for wearable computers.

An always-limited computational power, restricted interaction capabilities, and the constraint of mental user capacities are further properties of the Wearable Computing Paradigm that make solutions challenging as a result of those user interface limitations.

There are different approaches when defining wearable computing depending on the research direction and the application domain. It all starts with a remark from Steve Mann [3].

'Can you imagine hauling around a large, light-tight wooden trunk containing a co-worker or an assistant whom you take out only for occasional, brief interaction? For each session, you would have to open the box, wake up (boot) the assistant, and afterward seal him back in the box. Human dynamics aside, wouldn't that person seem like more of a burden than a help? In some ways, today's multimedia portables are just as burdensome.'

2.1   Interaction in Wearable Computing

The interaction between the user, the system and the environment is used to explain the difference between pervasive and wearable computing. In pervasive and even mobile computing, the information display and the information input device are modified desktop human computer interfaces (HCI).

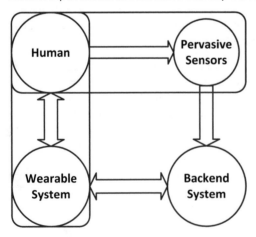

Figure 1: Wearable computing interaction paradigm

To interact, the user needs to focus on the display and the input device. This catches the user's attention as well as his physical activity, especially the use of the hands. Consequently, the user can interact either with the system or with the environment but not with both of them at the same time. This means that the full attention of the user is needed for engagement in the serious game. As soon as the user interacts with the environment, the user loses the contact to the game.

In contrast, wearable systems are permanently useful and usable. They allow simultaneous interaction with the system and the environment by using sensors for direct interaction between the system and the environment as well as the capability of the system to mediate the interaction between the user and the environment (figure 1).

2.2   Challenging Wearable Computing Systems

A Wearable Computing system is specific: It has its own requirements and challenges for the developer during implementation due to the specific interaction concept and environmental setting of the user. As such, any Wearable Computing system requires a Pervasive Computing setting providing the necessary communication and computational infrastructure. The implementation of the wearable interaction concept involves four main issues:

The system must be able to interact with the environment through an array of different sensors distributed in different parts of the outfit. In particular, it must be able to develop a certain degree of awareness of the user's activity, their physiological and emotional state, and the situation around the user. This is often referred to as context awareness.

The user interface should require minimal cognitive effort with no or little involvement of the hands. In general, one achieves the low cognitive load through appropriate use of the context information. Thus, for example, instead of having the user select a function from a complex hierarchy of menus, the system should derive the two most likely options from the context information and present the user with a simple binary choice. In terms of the actual input modality, simple, natural methods such as nod of the head, a simple gesture, or spoken commands are preferred.

Using context information, the system should be able to perform a wide range of tasks without any user interaction at all. This includes system self-configuration tasks as well as automatic retrieval, delivery, and recording of information that might be relevant to the user in a specific situation. A trivial example of a context-dependent reconfiguration could be a mobile phone that automatically switches off the ringer during a meeting.

Seamless integration in the wearable-outfit is necessary so that it neither interferes with the user's physical activity nor affects their appearance in any undesirable way.

To better understand the actual development, we summarise the development of computing and communication technology, look at the origins of wearable computing and current results when seeking to identify rehabilitation applications with specific user needs.

2.3   Wearable Computing Evolution

The evolution of electronics for computing and communication hardware and the use and users of this hardware also influenced the development of Wearable Computing. Between 1960 and 1980, computer hardware became more powerful, smaller and cheaper. Just a little reminder: In 1980 we got the personal computer, the Gameboy from Nintendo came in 1989 into the market and GSM mobile phones came in 1990.

As can be seen from these hardware development examples, the usage and the users of computer hardware also changed quite a bit during those decades. In the 1950s, experts used to build computers in order to solve their problems. In the 1960s, computer companies built computers for a commercial use for the first time. In the 1980s, the personal (desktop) computer was for geek users. People also used these desktop computers for games and, with the introduction of the TCP/IP protocol and the establishment of the DARPA Net, for communication purposes. It was in the 1990s when, with the introduction of the first browser, the internet started to grow. The target group became 'everyone'. In the 2000s with the availability of WLAN, GPRS and UMTS 'everywhere' became reality. Moreover, with the short-range communication standard of Bluetooth® in 1999, the Personal Area Network concept arose.

However, it was not only the user that evolved with the hardware, it was also the use of it. The early computers (until 1960s) could only solve specific problems of number crunching and the organization of large amounts of data. Later (until the 1980s) they were used to solve technical problems and in business support like accounting. In the 1990s, we saw the internet and communication (email, chat etc.); in 2000s, mobile devices became popular for casual use like SMS, WAP etc.

The wearable computing idea emerged in the second half of the 1990s with research focusing on the idea to miniaturize the Personal Computer even further and find a solution for its use in mobile situations, addressing the following properties and constraints:

- **Limited Capabilities:** The wearable computer system is usually limited or constrained in terms of available computation power, energy consumption, and available I/O modalities compared to the capabilities of stationary computer systems [4].
- **Operation Constancy:** The wearable computer system should be always on to provide the user with useful function. Here, the user's primary task is the main focus, requiring the wearable system to be a secondary support in the background [4], [5], [6].
- **Context-Awareness:** The wearable computer system senses properties (of the environment or the user) useful to support the user during a primary task and supposed to optimize interaction [4], [5], [7].
- **Seamless Environment Integration:** The wearable computer system has to be unobtrusive and non-distracting for the user during a primary physical task. Any specific context of use requires a tailored user interface [8].
- **Adapted Interaction:** The wearable computer system may automatically adapt the interaction style and/or interface rendering of a running application in order to make interaction easier and more efficient while minimizing the mental effort [4], [7], [9], [10].

In recent years, the projects involving our research teams focused on context awareness and adapted interaction. The limited capabilities are evident even though the miniaturization and Moore's law constantly provide continuously increasing computing power. The seamless environment integration becomes more and more of a reality thanks the driving force of pervasive and mobile computing. The need of operation constancy is, in the first instance, a challenge for energy management but hawse had to address this by specific concepts of context awareness and adapted interaction, taking into account that saving energy is one of the needs the context of the application requires.

2.4   Wearable Computing in Healthcare

The research described here deals with three application domains that all tried to use as many commercial off-the-shelf hardware components as possible: (1) applications in the hospital, (2) applications supporting or monitoring people with handicaps, chronic diseases or the wish of greater mobility, and (3) applications in rehabilitation.

wearIT@work

In the wearIT@work research project (www.wearITatwork.com) [11] we had two applications in healthcare. The first one was supporting the physician and the nurse during a ward round by accessing the patients' clinical data and the hospital information system (HIS) for checking and ordering treatments. The second was helping visually impaired people indoors and outdoors by the use of a wearable navigation system.

The basic concept behind the project at that time (2004-2009) was to use available commercial off-the-shelf hardware components as far as possible. Only the physician and the visually impaired used a wristband and data glove for interaction in the two healthcare applications.[5]

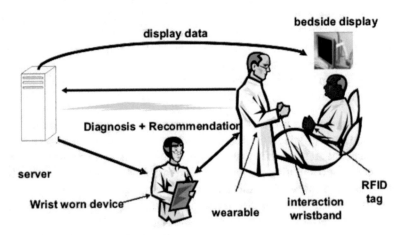

Figure 2 wearIT@work healthcare solution

The main research in the project was on an Open Wearable Computing Hardware Platform that found an implementation later in the Google Glasses and in many other Head Mounted Computers which have been on the market since 2012. The other was an Open Wearable Computing Framework that al-

---

[5] http://www.wearitatwork.com/available-technologies-and-their-use-in-wearitwork-project/hardware/io-devices/clothing-io-devices/; accessed 21.3.2016

lowed application developers to focus on the specific application and use readily available modules e.g. for speech, collaboration, communication and workflow support [11].

The solution for the ward round was the merger of a wearable and a pervasive computing solution where the bedside (computer) screen, usually used for patient entertainment , allowed the physician via WLAN to access the Hospital Information System – HIS (see figure 2). The patient had a wristband with an RFID tag. The physician had a headset for speech input and a wristband with an RFID reader and a motion sensor. The nurse had an RFID tag and a PDA with WLAN access to the HIS.

The end-users (physicians and nurses) and the management (CIO and CFO) were very positive as the solution reduced media conversations, e.g. from spoken word to text, increased the data quality, and used the existing IT infrastructure. Pure organizational issues were also essential for a successful introduction of the solution. An example was the replacement of the signature of the physician on a medication by the proximity of the physician and the nurse at the patient's bed. However, in the end, the gesture recognition was for some of the users not robust enough and therefore not acceptable in this environment.

uWear

The navigation solution for the visually impaired (see figure 3) used the lightweight OQO computer (https://en.wikipedia.org/wiki/OQO), a standard GPS receiver, the above-mentioned data-glove, specialized speech recognition software, and a WLAN based localization scheme for the indoor scenario. The user received instructions through a speaker integrated into a scarf. The results were very promising. However, robustness, missing detailed maps with personalized instructions and configurability of the overall system preventing us from going beyond the research prototype. In addition, there was a lack of business potential.

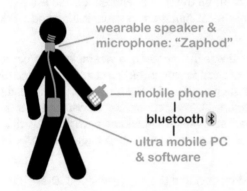

wearable speaker & microphone: "Zaphod"

mobile phone

bluetooth ✳

ultra mobile PC & software

Figure 3 uWear navigation system for visually impaired

Wearable Computing for People with Chronic Diseases

The next group of projects were those with solutions for people with chronic diseases: CHRONIOUS (www.chronious.eu), USEFIL (www.usefil.eu), Help-AAL (www.help-project-parkinson.com/) and REMPARK (www.rempark.eu). The common characteristic of these projects is that that they are designed to support people with the need for continuous medication and monitoring of the disease's progression. At the start of these projects we observed that those people repeatedly spent time in hospitals for observation and adjustment of the medication. On the one hand we observed that these patients wished to avoid these hospital stays, which were always triggered by a deterioration in the patient's health status. On the other hand the physicians were wanting to be able to better monitor the patient's health status in order to to intervene as early as possible by adjusting the medication or recommending a behavioural change.

CHRONIOUS

CHRONIOUS (2008-2012) provided a smart wearable platform for monitoring people suffering from Chronic Obstructive Pulmonary Disease (COPD) and Chronic Kidney Disease (CKD) and Renal Insufficiency. The core element was a T-shirt with integrated sensors and additional sensors placed in the living environment of the patient at home. The system generates alerts if health and behavioural data lay outside well-established patterns. A multi-parametric expert system analyses the collected data and notifies the healthcare provider in critical cases through a dedicated and secure web platform. The integrated platform (figure 4) provides real-time patient monitoring and supervision, both indoors and outdoors and represents a generic platform for the management of various chronic diseases. In addition, an ontological information retrieval system satisfies the need for up-to-date clinical information.

CHRONIOUS validated the platform with 100 patients in clinical trials in medical centers and patient's home environments. The focus was to provide the physician with reliable data on the patients' health status. The feedback for the patient to self-monitor was not a focus. The main outcome was the development of the platform as a model for future generations of 'chronic disease management systems'. The usability and user experience of the home patient monitor left room for improvement as a focal point in the next project. It was the first time that we experienced the need for pervasive and wearable computing systems installed in the patient home with occupants who were not experienced in the use of such IT systems. We had motivated patients since staying at home is always preferable to a stay in hospital. However, in when a user cannot navigate the monitoring system, their frustration changes their perspective on the usability of the solution. That is why we see the urgent need to focus more on user acceptance than the technical solution.

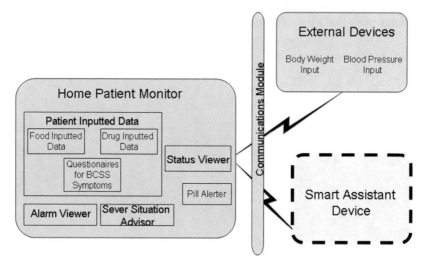

Figure 4 CHRONIOUS system overview

Help-AAL and REMPARK

The projects Help-AAL (2009-2011) and REMPARK (2011-2015) were both Joint-AAL projects with a stronger focus on the commercial exploitation of the solutions, in this case for patients with Parkinson disease. Both projects used wearable and pervasive computing technology to monitor patients, better understand their needs, and provide medication when the sensors indicated the need for this. There were only minor hardware developments needed; the focus was on the applicability of the solution. The patients' urgent need for help led to sufficient user acceptance.

The Personal Health System (PHS) of REMPARK (figure 5) has closed loop detection, response and treatment capabilities for management of Parkinson's Disease (PD) patients using a wearable monitoring system to identify in real time the motor status of the PD patients, and evaluation of ON/OFF/Dyskinesia status in operation during ambulatory conditions. The gait guidance system helps the patients in real time to carry out their daily activities. The analysis of the data provided supports the neurologist in charge with a disease management system accessing accurate and reliable information to decide the most suitable treatment for the patient. The evaluation used tests with 60 patients from four medical centers [12].

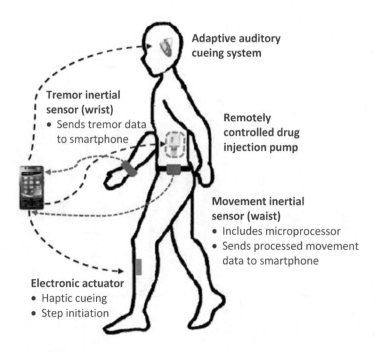

Adaptive auditory
cueing system

Tremor inertial
sensor (wrist)
• Sends tremor data
  to smartphone

Remotely
controlled drug
injection pump

Movement inertial
sensor (waist)
• Includes microprocessor
• Sends processed movement
  data to smartphone

Electronic actuator
• Haptic cueing
• Step initiation

Figure 5 REMPARK system overview

USEFIL

Creating user acceptance necessitaes bridging the gap between technological research advances and the practical needs of elderly people. This means, in practice, developing advanced but affordable, unobtrusive in-home monitoring and web communication solutions. In the USEFIL project, commercial off-the-shelf technology for applicable services assisted the elderly in maintaining their independence and daily activities. The installation of the system had to avoid any retrofitting in the person's home and needed to be unobtrusive once installed. The key aspect was that the technology implementation has the user's acceptance and user interactions truly address user needs. The system as shown in figure 6 provides access to care services and uses web technologies for information sharing and social services.

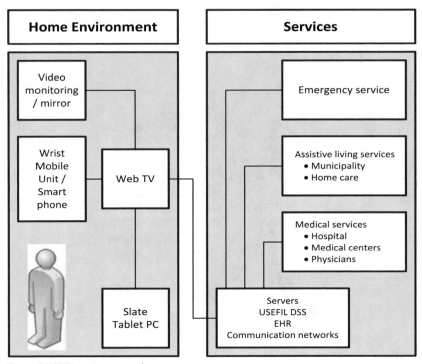

Figure 6 USEFIL system overview

There was no specific chronic disease addressed in the applications. The knowledge that, with aging, multi-morbidity and the probability of a chronic disease was sufficient motivation for the project. However, even though the technology was all off-the-shelf, overcoming the fear of contact and misuse, achieving user acceptance, and motivating the ongoing use of the system brought up the idea of using Serious Games for motivation purposes and for teaching how to use technology that was relatively new for all of the users.

Another aspect became obvious in the project: there is not 'the' elderly person: With aging, the probability of mild and moderate vision or hearing impairments and movement disorders increases. As designers usually do not have their own experiences of these problems, most off-the-shelf products are not designed to meet the requirements of such people.

The concept of personas should therefore be included into the design process; modelling virtual users with typical impairments and disorders is necessary. Those personas have to perform in virtual reality the interaction tasks with the product and the environment as part of the design process. This leads to an extension of the design model to one similar to integrating virtual mannequins in crash tests. Within the VICON project, we applied this methodology to the development process of white and brown goods[6] using a washing machine, a remote control and a mobile phone as examples of off-the-shelf products.

3 Understanding the User

The aim of any system design is to successfully integrate a broad range of diverse human factors into the development process with the intention of making systems accessible to and usable by the largest possible group of users. In wearable and mobile computing, one has to include in the system design process the hardware and interaction design. This requires an extension of the classical user centred design (UCD) approach.

Although manufacturers are nowadays more likely to invest efforts in user studies, products in general only nominally fulfil, if at all, the accessibility requirements of as many users as they potentially could. The main reason is that any user-centred design prototyping or testing aimed at incorporating real user input comes at a rather late stage of the product development process. Thus, the more progressed a product design has evolved,- the more time-consuming and costly it will be to alter the design.

The number of functions and features requiring user attention and interaction has increased significantly in wearable computing solutions and also mobile phones, remote controls, or even washing machines and dish washers. For people e.g. with age related mild-to-moderate physical or sensory impairments, this creates difficulties and rejection of those (sub-) systems. To anticipate and avoid this requires the acknowledgment of needs as early as possible in the development process.

Typical use-cases include a wider range of environments. The impact of factors such as location, mobility, and social and physical environments increases the level of comprehension and attention needed to operate the system. Accordingly, the features and capabilities have to take the context of use into account. This requires supportive functions in design tools such as Computer

---

[6] electrical appliances and electronic consumer durables

Aided Design (CAD). These functions have to promote the context-related design of products by default, enabling designers to understand errors in the design and recommend how to fix them. The design process has three phases: the sketch phase, the CAD phase and the evaluation phase. This design process runs in a cyclical fashion as specified in ISO 9241-210:2010.

3.1   Virtual User Model Methodology

This methodology requires a Context Model with the capability to determine recommendations for appropriate interaction design [13]. Therefore, it incorporates well-defined partial models, which are logically interrelated to each other in order to determine appropriate recommendations for the designer.

Using the results of observational studies ([14] and [15]) one can elaborate a suitable taxonomy for the context model (respectively virtual user model), which consists of the following partial models:

User Model, where all information about the potential users of the product is stored. Focus is upon exemplary users with e.g. mild to moderate physical impairments. One divides the respective user models into several subgroups (profiles), and divides them e.g. into different levels of impairments. Additionally one has mixed profiles describing the potential users who are subject e.g. to a mixture of hearing, sight and dexterity impairments.

Component Model describes the user interface components and adds functionalities to specific instances, e.g. a button. The button consists of the functional attribute of a switch with two states. This model is used to connect recommendations with components - especially in the CAD phase.

Model for Recommendations In this model, guidelines and experience of the designer are stored. These consist of the predicates 'Name', 'Text', 'Summary', Rules, Phases and an Attachment, where e.g. Sketch Phase Template Layers can be stored. A component attribute defines rule sets for the design phase if a recommendation relates to a specific component or component functionality like 'Audio Output".

Environment Model, where all data of the environment is stored. That includes the physical conditions of the environment of the real world, objects and characteristics of the environment etc.

Task Model, which describes how to perform activities to reach a predefined goal. This model may be based e.g. on Hierarchical Task Analysis (HTA) providing an interface where the designer can define actions of the user for the evaluation in the virtual environment envisaged in the evaluation phase.

The Overall Architecture

In relation to functional requirements, such as gaining component recommendations as an output, the virtual user model (VUM) needs to be able to parse the sub-models using logical constraints. This is necessary in order to build an inference model with all relevant data.

For the implementation, an architecture is proposed which includes the VUM as a knowledge base. Figure 7 shows the VICON system architecture with its different parts.

All data of the VUM is stored In the backend.. The frontend integrates company-specific design and testing applications as well as all client-specific features to obtain recommendations (recommendation module). The middleware layer provides a seamlessly accessible connection between the frontend applications and the reasoning engine with a socket connection handler and socket server.

All recommendations are marked with a phase attribute which defines at which phase a recommendation is relevant. Additionally, every recommendation instance consists of a user model-, environment-, task- or component rule. The backend services provide the access to the ontology-schemes, algorithms and data in order to control and manage the framework.

The sketch phase provides qualitative design recommendations to the designer. The designer can use these recommendations for drafting the user interfaces. In order to offer the designer flexibility within the creativity process, the tool offers only qualitative (high level) design recommendations.

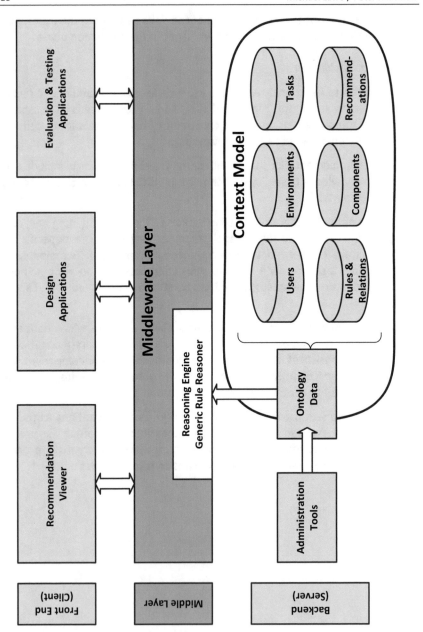

Figure 7 VICON software framework architecture

Setting up the Virtual User Model in the Sketch Phase

An idealized workflow could mean having corresponding recommendations like 'use a maximum of 5 buttons in total' directly on screen while drafting the product-shape on paper or a digitizer tablet. The designer can save all the settings amongst the given recommendations to 'import' all information into the subsequent phase related modules.

The idea is that the Virtual User Model establishes steps which apply specified rule sets by every step as illustrated in figure 8.

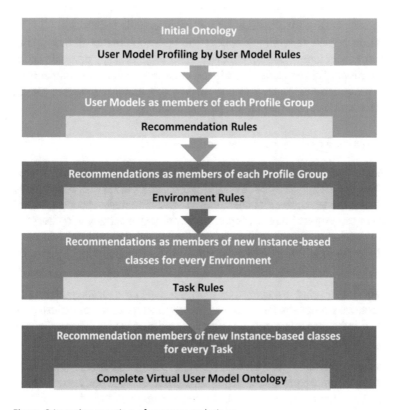

Figure 8 Iterative creation of recommendations

These steps are:

- **Applying user model rules:** The General Rule Reasoner uses the user model rules to define all instances of the user model class as members of specified WHO ICF profiles (e.g. a specific profile for moderate hearing-impaired people).

- **Generation of initial recommendations:** This step is the same as the first step, with the difference that it uses the recommendation rules and instances based upon user model profiles.
- **Creation of environment recommendations:** This step creates classes which are based on the id names (IDs) of every environment and adds all textual and component recommendations which were reasoned by the environment rules as members of these new recommendation classes (e.g. a recommendation class for an instance of the environment). These rules can also use the previous defined recommendation classes.
- **Creation of task recommendations:** The last step creates all task related recommendations based on task rules and all previously defined recommendations. This procedure is the same as the creation of environment recommendations. All task id names define dynamically created classes which contain recommendations for specific tasks.

3.2   Discussion of the Virtual User Model Approach

The more tools for design are used by product designers, the less their impact will be upon the design process. In other words, the earlier a product meets the user's requirements, the less changes will affect the design process, including effects in the project budget, project plan and design activity. It is therefore of crucial importance to understand the product development process in detail in order to know in which phases to integrate tools that support inclusive design.

To establish tools that are able to be integrated into the existing design process is therefore one of the key issues in fostering acceptance of inclusive design tools in industry. This offers an effective and non-disruptive way to present design recommendations to the product developer.

The presented design approach based on the described context model is capable of supporting the product development process in the early stage before the realization of prototypes. It should however not be regarded as a total substitute for real users, but more as complementing the involvement of real users and an opportunity to minimize the effort of applying more costly and time-consuming techniques in the early design phase.

In the evaluation phase, designers have the ability to test their recommendation-based product design in a virtual environment through a digital human model which corresponds to the data of the Virtual User Model as configured by the designer in the sketch phase.

The proposed design approach envisages that the output of the evaluation phase flows back into the preceding phases such as the CAD phase and sketch phase. In this way, a continuous update of the context model can be created.

The focus of the research was mainly upon the feasibility of the approach. To guarantee the validity of the data in the context model requires as much data as possible. We conducted a comprehensive field study within this research. The quality of the data (recommendations, design constraints, UI components, etc.) in the context model is highly dependent upon the availability of quantitative, accurate data. A comprehensive database will change the design process in wearable and pervasive computing quite drastically.

The main outcome of this research is published in the following reference [16] but more detailed insights provide [17], [18] and [19].

4 Serious Games and the Usefulness of Exergames

During the past decade, Serious Games have emerged as a significant research area which combines game-based concepts with the ICT related domains, resulting in the broad spectrum of applications. Such applications vary from military training [20] to education [21], healthcare [22], [23] and a myriad of others. There are numerous definitions of the term 'serious game' summarized by Rego et al. [24]. However, they are consistent in their reference to the use of computer games beyond the purpose of entertainment. In other words, Serious Games is a game-based approach that aims to achieve certain high-level goals rather than providing pure fun, while exploiting the entertainment mechanisms to stimulate the user's engagement. In particular, the rehabilitation process can greatly benefit from the concept of Serious Games since the traditional treatment approaches are typically repetitive and not appealing for patients, resulting in low compliance to the prescribed therapy sessions and frequent dropouts. On the other hand, the fusion of game-based contexts and the ICT-based approaches can benefit not only by generating motivation in rehabilitation sessions and patient self-confidence, but also by potentially providing more precise assessments of the patients' performance (in comparison to human observers) and thereby provide support for a more correct execution of the exercises and a long term monitoring of patients' progress. Another important aspect is that games distract the patient's attention and, as such, can be used to aid in the management of pain as Rego et al. report.

The Serious Games concepts used for rehabilitation purposes differ in a number of aspects to the interaction designs and technologies used depending on the kind of rehabilitation such as disease, injury, or mental skills and targeted skills. Evaluating the state of the art of rehabilitation systems based on game-context it is possible to identify different types of approach depending on the aim of rehabilitation: Beyond entertainment, games target enhancing cognitive, sensory-motor, emotional, personal and social competences. [25]

However, games mostly require cognitive and motor activity and are thus mostly used for rehabilitation of cognitive and physical or motor skills. Amongst others, cognitive rehabilitation includes regaining skills related to memory, attention and concentration, reasoning and problem solving, sentience, language, volition and judgment. On the other hand, motor rehabilitation includes exercises for recovering from effects of neurological or physiological damage. Here they are used for post-stroke, balance training, mobility, Parkinson's disease, orthopaedic rehabilitation, and so on.

Concerning interaction, we distinguish between input and output methods for the interaction between the system and the patient. The input technologies relevant for the rehabilitation systems include standard interfaces such as keyboard and mouse, and audio and video systems such as microphones and web camera systems. For motion tracking, devices such as Nintendo Wii, Microsoft Kinect and different kinds of gloves with embedded sensors, haptics etc. exist as sensor based solutions. As for the output technology, the current works mostly rely on desktop monitors, TVs, and virtual reality (VR) equipment like head-mounted displays.

In the game design, we have to refer to all the characteristics and features of the game, which are relevant for the patients' engagement in the rehabilitation sessions. In particular, whether these are single or multi-player modes, where, with the later social factors, can be an additional motivational driver. Another aspect is the game genre. Depending on the technology used and the kind of rehabilitation, one can consider simple motion-based games with the focus solely on the quality of movement or on the other hand, simulations, strategy and other kinds of games focused on the quality of both cognitive reasoning and movements.

Essential for Serious Games more than other types of game is automatic performance assessment. The automatic assessment of performance represents, for rehabilitation systems, the measurement of the therapeutic effect. By identifying correct and incorrect actions, one enables an adequate interaction with the game, adapting the response. Here the option is to switch to easier or more difficult levels of the game. Games in rehabilitation have to give feedback to patients as a motivation for performing exercises and to doctors for providing information about the progress of the rehabilitation.

Rehabilitation systems based on serious game concepts can be used either at home or in hospital settings.

4.1   State of the Art

In the following, we provide a brief review of recent research on designing innovative and advanced solutions for neuro-cognitive and motor rehabilitation that have inspired our studies.

With a focus on mechanisms for adapting to the patients' condition, one rehabilitation system prototype [26] integrates video games with computational intelligence for both online monitoring of the movements' execution during the games and for adapting the gameplay to the patients' status. The proposed system applies real-time adaptation to modify the gameplay in response to the current patient's performance and progress and/or to the exercise plan specified by the therapist.

A system for encouraging stroke patients to perform physical exercises targeting upper limb motor rehabilitation [27] requires physiotherapists to set the rehabilitation parameters reflected in the games tasks of reaching, grasping and caching virtual objects in the virtual environment through 'real-life" physical movements. The input devices include gloves for recognizing capture finger flex and hand positions, and wireless magnetic sensors for tracking hands, arms and upper body movements. The output has visual and audio modalities using a desktop computer for the operating clinicians or therapists and a head mounted display for the patient.

Several games use design principles for upper limb stroke rehabilitation [28]. Input is provided by low cost webcams that track patients' movements, while output is done through a PC display. By analysing the log files, it is possible to provide informative parameters related to the patients' performance – a color-coded mapping of the movement trajectories with the intensity of the colours used to represent the frequency of the players' hand movements in certain areas. This approach is similar to a 'heat map". Other parameters extracted include the score related to 'hits" and 'misses" in target areas, progress over time, and the amount of time spent playing each game.

There is a Serious Games approach for people with behavioural and addictive disorders like eating disorders and pathological gambling [29]. The prototype developed involves the player in an interactive scenario designed to increase the awareness of the general problem, introducing solving strategies, self-control skills and control over impulsive behaviour. The system supports multi-parametric input including speech, touch, biosensors and motion tracking.

4.2   The Need for Further Research

There were a couple of research projects in the field of Serious Games for rehabilitation.

The REWIRE project[7] develops an integrated low-cost system for home rehabilitation with an adaptive game engine. It supports the integration of natural user interfaces, a wide variety of game scenarios, quantitative and qualitative exercise evaluation, and audio-visual feedback. The whole system consists of a patient station mounted at home and a hospital station and networking station at the health provider site. The therapist can customize the game interaction mechanisms and select the proper device to maximize the rehabilitation efficacy from a range of supported devices including Sony PlayStation 3 Eye, Microsoft Kinect, Wii Balance Board, Omni Phantom and Novint Falcon.

The RehaCom system[8] is for cognitive rehabilitation and is well established in many hospitals and clinics. The first version dates back to 1986 and since then it has been refined over 30 years in clinics. The system consists of training procedures intended for different skills, namely attention, executive, memory, field of view and visuo-motor skills (ability to synchronize visual information with physical movement).

Duckneglect[9],is a low cost platform based on video-games, targeted to neglected rehabilitation. The patient is guided to explore his neglected hemispace by a set of specifically designed games that ask him to reach targets with an increasing level of difficulty. Visual and auditory cues help the patient in the task and are progressively removed. A controlled randomization of scenarios, targets and distractors, a balanced reward system and music played in the

---

[7]   https://sites.google.com/site/projectrewire/home accessed 30.3.2016
[8]   https://www.hasomed.de/en/home.html accessed 30.3.2016
[9]   http://www.ncbi.nlm.nih.gov/pubmed/23510971 accessed 30.3.2016

background, all contribute to make rehabilitation more attractive, thus enabling intensive prolonged treatment.

There is much more available and these will be mentioned in the subsequent book chapter. However, the current approaches typically remain limited with respect to the following matters:

- Lacking the attractiveness of commercial games, due to limited graphics and game-play; – thus they are prone to become boring to the patient after a short period.
- Requiring expert knowledge (either technical or medical) for setting up the equipment properly;
- Providing limited opportunities to adapt to specific patient needs;
- Targeting one specific type of rehabilitation (stroke, cognitive stimulation, etc.);
- Lack of details on possible costs and portability issues involved with the deployment of the solutions proposed in real life settings.

That is why there is a need for a serious games approach for rehabilitation using wearable and pervasive computing. There have been major developments from the early days of the technology to the commercial off-the-shelf situation for most wearable devices like head mounted displays and wrist-worn devices. The products have reached the consumer and have the potential of being beneficial for use in rehabilitation and Serious Games.

5 Further Outline

In the following book chapters, we will first introduce in more detail the Rehab@Home project that initiated the idea of this book. We will introduce the concept of a Reference Rehabilitation Platform (chapter two).

The main issues of a comprehensive and flexible hardware and software design for such an application domain we outline in chapter three. It is a matter of fact that technology develops continuously, making it hard for developers to offer a solution that is more than a snapshot of the development at the current time. The approach presented here takes account of this fact, offers an open source approach and allows later developments to reuse as much as possible our results. However, those readers more interested in the application than the technology might skip reading chapters two and three.

Serious Games are a mechanism to motivate the users to perform essential exercises with the engagement needed. They intend to create and sustain the patient's motivation for this purpose. However, especially after a stroke, patients are desperate and have no lack of motivation. The requirements elicitation process is a challenge for Serious Games development. In the context of rehabilitation, it is described in chapter four.

When using wearable and pervasive computing solutions one collects a huge amount of data. The technical concept of health data collection and integration is the topic of chapter five.

There are significant ethical, legal, and social implications to observe in homecare. The influence on our topic we outline in chapter six, when looking at media ecology aspects.

A very specific experience is research into the domain of mild to moderate cognitive impairments; in chapter seven we report on our insights concerning this when intending to evaluate technical solutions.

Any medical solution requires clinical tests. This is a challenge when the solution aims at home-rehabilitation. In chapter eight, we describe our experiences concerning past and future clinical tests.

Motivation is a key issue when using games and Serious Games, particularly for rehabilitation purposes. Only motivated persons can benefit of such games. Having reliable information about the emotional state of a person using such games is required. We describe an approach to detect the emotional state by using physiological data collected by sensors which are also used for measuring the health status of the person. Our attempt to research approaches to monitor emotional state is the topic of chapter nine.

6   References

[1]   ABT, Clark C. Serious games. University Press of America, 1987.

[2]   PENG, Wei; LIN, Jih-Hsuan; CROUSE, Julia. Is playing exergames really exercising? A meta-analysis of energy expenditure in active video games. Cyberpsychology, Behavior, and Social Networking, 2011, 14. Jg., Nr. 11, pp. 681-688.

[3]   MANN, Steve. Wearable computing: A first step toward personal imaging. Computer, 1997, 30. Jg., Nr. 2, pp. 25-32.

[4]   Starner, T. (1999). Wearable Computing and Contextual Awareness. MIT Boston: PhD thesis

[5]   Rhodes, B. (1997, March). The Wearable Rememberance Agent. 1 (4), pp. 218-225.

[6]   Mann, S. (1997). A Historical Account of 'wearcomp' and 'wearcam' Inventions Developed for Applications in 'personal imaging'. ISWC '97 Proceedings of the 1st IEEE International Symposium on Wearable Computers, pp. 66-73.

[7]   Kortuem, G., Segall, Z., & Bauer, M. (1998). Context-aware, Adaptive Wearable Computers as Remote Interfaces to 'Intelligent' Environments. ISWC '98 Proceedings of the 2nd IEEE International Symposium on Wearable Computers, (p. 58). Washington DC, USA.

[8]   Starner, T. (2002). Attention, Memory, and Wearable Interfaces. IEEE Pervasive Computing, 1 (4), pp. 88-91.

[9]   Schmidt, A. (2002). Ubiquitous Computing - Computing in Context. Lancaster University, UK: PhD thesis.

[10]   Schmidt, A., Gellersen, H., Beigl, M., & Thate, O. (2000). Developing User Interfaces for Wearable Computers: Don't stop to point and click. IMC Proceedings of the International Workshop on Interactive Applications of Mobile Computing.

[11]   PEZZLO, Rachel; PASHER, Edna; LAWO, Michael (Eds.) Intelligent Clothing: Empowering the Mobile Worker by Wearable Computing, Berlin: Aka / IOS Press, 2009.

[12]   Ahlrichs, C. (2015) Development and Evaluation of AI-based Parkinson's Disease Related Motor Symptom Detection Algorithms. University Bremen/Germany: PhD thesis.

[13]   Bürgy, C. and J. Garett (2002). Situation Aware Interface Design: An Inter-action Constraints Model for Finding the Right Interaction for Mobile and Wearable Computer Systems. 19th International Symposium on Automation and Robotics in Construction Gaithersburg, Maryland: pp.563-568.

[14]   Castillo, E., J. M. Gutiérrez, et al. (1997). Expert systems and probabilistic network models, Springer Verlag.

[15]   Clarkson, J., Coleman, R., Hosking, I., Waller, S. (2007). Inclusive Design Toolkit, Cambridge Engineering Design Centre.

[16]   MATIOUK, Svetlana, et al. Prototype of a virtual user modeling software framework for inclusive design of consumer products and user interfaces. In: Universal Access in Human-Computer Interaction. Design Methods, Tools, and Interaction Techniques for eInclusion. Springer Berlin Heidelberg, 2013. pp. 59-66.

[17]   Modzelewski, M. (2014). An ontology-based approach to achieve inclusive design support in the early phases of the product development process. University Bremen/Germany: PhD Thesis.

[18]   KIRISCI, Pierre T. Gestaltung mobiler Interaktionsgeräte. ISBN 978-3-658-13247-7 (eBook), 2014.

[19]   Kirisci, P.; Thoben, K.-D.; Klein, P.; Hilbig, M.; Modzelewski, M.; Lawo, M.; Fennell, A.; O'Connor, J.; Fiddian, T.; Mohamad, Y.; Klann, M.; Bergdahl, T.; Gokmen, H.; Klen, E.: Supporting inclusive design of mobile devices with a context model. Edited by Karahoca, A.: Advances and Applications in Mobile Computing, INTECH, 2012 ISBN: 978-953-51-0432-2, pp. 65-88

[20]   NUMRICH, Susan K. Culture, Models, and Games: Incorporating Warfare's Human Dimension. IEEE Intelligent Systems, 2008, Nr. 4, pp. 58-61.

[21]   VON WANGENHEIM, Christiane Gresse; SHULL, Forrest. To game or not to game? Software, IEEE, 2009, 26. Jg., Nr. 2, pp. 92-94.

[22]   MACEDONIA, Mike. Virtual worlds: a new reality for treating posttraumatic stress disorder. Computer Graphics and Applications, IEEE, 2009, 29. Jg., Nr. 1, pp. 86-88.

[23]   SAWYER, Ben. From cells to cell processors: the integration of health and video games. Computer Graphics and Applications, IEEE, 2008, 28. Jg., Nr. 6, pp. 83-85.

[24]   REGO, Paula; MOREIRA, Pedro Miguel; REIS, Luis Paulo. Serious games for rehabilitation: A survey and a classification towards a taxonomy. In: Information Systems and Technologies (CISTI), 2010 5th Iberian Conference on. IEEE, 2010. pp. 1-6.

[25]   WIEMEYER, Josef; KLIEM, Annika. Serious games in prevention and rehabilitation—a new panacea for elderly people? European Review of Aging and Physical Activity, 2012, 9. Jg., Nr. 1, pp. 41-50.

[26]   PIROVANO, Michele, et al. Self-adaptive games for rehabilitation at home. In: Computational Intelligence and Games (CIG), 2012 IEEE Conference on. IEEE, 2012. pp. 179-186.

[27]   MA, Minhua; BECHKOUM, Kamal. Serious games for movement therapy after stroke. In: Systems, Man and Cybernetics, 2008. SMC 2008. IEEE International Conference on. IEEE, 2008. pp. 1872-1877.

[28]   BURKE, James William, et al. Optimising engagement for stroke rehabilitation using Serious Games. The Visual Computer, 2009, 25. Jg., Nr. 12, pp. 1085-1099.

[29]   CONCONI, Alex, et al. Playmancer: A serious gaming 3d environment. In: Automated solutions for Cross Media Content and Multi-channel

Distribution, 2008. AXMEDIS'08. International Conference on. IEEE, 2008. pp. 111-117.

Acknowledgment

The authors of this chapter thank all their co-workers and national and European funding agencies of the many projects in recent years. Without their support this chapter could not exist.

II   Reference Rehabilitation Platform for Serious Games
The Rehab@Home Project

Peter Knackfuß / Michael Lawo

Abstract

In this book chapter, we describe a Reference Rehabilitation Platform (RRP) for Serious Games as developed in the three years EC funded eHealth project Rehab@Home. We refer to our experiences with its implementation and continuous evaluation and improvement. The RRP allows therapists, patients and caregivers to configure and monitor a sometimes – after a stroke or in case of chronic diseases like Multiple Sclerosis (MS) and Parkinson – a long-lasting rehabilitation process using Serious Games in the patient's home. The RRP supports both individual configuration to the patient's needs by the therapist and communication with the therapist and caregiver. We implemented the system in a cyclic user-centred design approach with three pilots and strong stakeholder involvement. Designed for the rehabilitation of the upper extremities, the RRP can be extended not only to the whole body and fine hand movements but also cognitive and aphasia training. Its modularity based on sensor integration enables this.

1	Introduction
2	Motivation
3	Objectives
4	Technology Description
5	Developments
6	Results
7	Business Benefits
8	Conclusion
9	References

© Springer Fachmedien Wiesbaden GmbH, part of Springer Nature 2018
M. Lawo und P. Knackfuß (Hrsg.), *Clinical Rehabilitation Experience Utilizing Serious Games*, Advanced Studies Mobile Research Center Bremen, https://doi.org/10.1007/978-3-658-21957-4_2

1   Introduction

Strokes or MS (multiple sclerosis) affect everybody differently. It is difficult to say how much of a recovery is possible and how long therapy will be necessary; sometimes it takes many years or requires a continuous therapeutic rehabilitation effort. The goal of rehabilitation is to help people to become as independent as possible and to attain the best possible quality of life. This is at its best in a familiar environment like the people's home. A cost effective and sleek infrastructure is beneficial in every case. A system should have integrated sensors able to collect relevant physical and medical parameters of a patients' status for check-ups and relapse prevention. It should support off- and online management and monitoring of the rehabilitation protocol, promote patient's social participation and community building. We assume that using Serious Games [1] is motivating patients for continuous rehabilitation efforts even when based at home and without the personal presence of a therapist.

It is obvious that this approach of using Serious Games is not something people can just use without any instruction and/or adjustment by a therapist [2]. For a sustainable approach it should be helpful to not only integrate patients and therapists but also caregivers or family members as an approach to enable improved motivation.

Under the assumption that a system useful for the target group of patients, therapists and caregivers exists, a further issue is to prove that the rehabilitation success is adequate. Tests with clinical partners over a longer period must prove the sustainability of the therapeutic approach using Serious Games by meeting the individual needs of the patient and reflecting the therapeutic intention. As the games are designed for use at the patient's home, only commercial off-the-shelf (COTS) interaction devices like the Microsoft KINECT, Nintendo Wii, LEAP-Motion Controller[1] or the no longer available Sifteo Cubes[2] were considered. Patients will use them together with COTS sensors like the Nonin Onyx II[3] to monitor physiological data. One unique selling point (USP) of the approach described and evaluated in this book chapter as an outcome of our research is the feature designed to easily calibrate the system by the therapist, thereby maintaining motivation through feedback from the system and

---

[1] https://www.leapmotion.com/ accessed 1.12.2015
[2] https://www.youtube.com/watch?v=dF0NOtctaME; accessed 1.12.2015
[3] http://www.nonin.com/Onyx9560; accessed 1.12.2015

the therapist when defining appropriate levels of activity for the individual patient as easily as possible.

Our findings are the outcome of the three years eHealth project Rehab@Home funded by the European Commission in the seventh framework programme under the contract no. 306113.

2 Motivation

Looking into the requirements for rehabilitation around mild cognitive impairments we argue that wearable and pervasive technology in combination with Serious Games is likely to be beneficial.

2.1 Rehabilitation

Stroke is the 2nd most common cause of death in Europe (1.24 million annual), in the European Union (508,000 annual), and the 3rd cause of death in Canada (14,000 annual) and the United States (over 143,000 people each year). Meanwhile, 1.8% of Asians aged 18 years and older have had a stroke. In general, according to the World Health Organization, about 15 million people suffer stroke worldwide each year. Of these, 5 million die and 10 million survive, albeit showing different degrees of disabilities.

Accordingly, the costs of stroke are enormous. Europe and the USA spent 2-6% of all health care costs on direct stroke care, inclusive of the costs of hospital and nursing home care, the services of physicians and other medical professionals, drugs, appliances, and rehabilitation [3]. Indirect costs, defined as production losses, further increase the burden of the disease. In Europe, direct costs are in the range of 3.000-16.000 € per patient during the first year, whereas the lifetime direct cost may reach 30.000 €. Taken together, direct and indirect costs may be as high as 20.000-26.000 € per patient in the first year. In Europe, 22 billion € are spent on stroke annually. [4]

Stroke affects everybody differently, and it is difficult to say how much of a recovery is possible. Many stroke survivors experience the most dramatic recovery during their stay in hospital in the weeks after their stroke. However, many stroke survivors continue to improve over a longer time, sometimes over a number of years. The goal of rehabilitation is to help survivors become as independent as possible and to attain the best possible quality of life. Rehabilitation does not 'cure' stroke in that it does not reverse brain damage. High quality rehabilitation, however, is essential to regain many – if not all – capabilities to lead a meaningful, fulfilling and even productive life.

The first stage of rehabilitation usually occurs within an acute-care hospital as soon as the patient is stable and the (initially high) risk of recurrence is lower. Ten percent of survivors can return home quickly but many need treatment in some type of medical facility. For over half of the stroke survivors, rehabilitation will be a long-term process requiring work with therapists and specialized equipment for months or (ideally) years after the stroke.

However, increasing cost pressure on the health system will lead to shorter periods of intensive rehabilitation at specialized facilities. Within this context, the adoption of suitable technical aids at home, together with a proper training based on a personalized program of exercises, can help to reduce the patient's stay at the hospital as well as the need for moving him/her forth and back to/from a physiotherapy unit or a paramedical structure.

Rehabilitation, which may be effective in improving the physical and mental condition of older people in long-term care, is a complex set of processes usually involving several professional disciplines and aimed at improving the quality of life of older people facing daily living difficulties caused by either temporary and/or chronic diseases. Comprehensive rehabilitation needs to address a number of different levels which may be contributing to loss of function: the damaged body part and other related body elements, psychological attitudes, immediate material environment (e.g. clothing items), the surrounding indoor environment (e.g. housing/equipment), external environment (e.g. shops, social outlets), social support networks.

In the specific case of stroke, although not exclusively, rehabilitation is based on neuroplasticity (also known as cortical re-mapping), which is the brain's ability to reorganize itself by forming new connections, allowing nerve cells in the brain to compensate for defects. However, neuroplasticity is only happening when there is the 'right' stimulus and sustainable motivation which are the key factors in successful rehabilitation. Rehabilitation teaches new ways of performing tasks to circumvent or compensate for any residual disabilities. There is a strong consensus among rehabilitation experts that the most important element in any rehabilitation program is carefully directed, well-focused, repetitive practice - the same kind of practice used by all people when they learn any new skill, such as playing guitar or skating.

Rehabilitative therapy begins in the acute-care hospital after stabilizing the patient's medical condition. The first steps involve promoting independent movement of the mostly paralyzed or seriously weakened patients. Patients are prompted to engage in passive (the therapist e.g. actively helps the patient move a limb repeatedly) or active (exercises are performed by the patient with no physical assistance) range-of-motion exercises to strengthen e.g. stroke-impaired limbs. Rehabilitation nurses and therapists help patients perform progressively more complex and demanding tasks and encourage patients to begin using their stroke-impaired limbs while engaging in those tasks. Beginning to reacquire the ability to carry out these basic activities of daily living represents the first stage of returning to functional independence.

Until Recently, it was assumed that early rehabilitation is the key to achieve a good functionality recovery for hands, arms or legs. Delaying the rehabilitation process leads to an expectation of no major improvements. Recently, Wallace et al. demonstrated that it is possible to obtain relevant results from rehabilitation many years after a stroke by applying a specific and very intensive rehabilitation approach [5]. This opens new horizons and puts the emphasis on the need for innovation in rehabilitation, both in terms of methodology and technology.

Having the cost of rehabilitation in mind, medium to long-term rehabilitation steps and activities could take place at the patient's home instead of at a dedicated treatment unit. The patient's home can become the place of physical and cognitive rehabilitation. This requires intensive and motivating training controlled by technology supported therapists and caregivers.

In practice, after a short rehabilitative period in a specialized centre, the patient can go home to continue rehabilitation with a set of easy-to-use technical enabling equipment to engage in a motivating personalized program. The patient will not only be in contact with experts in the rehabilitation centre – providing guidance and feedback – but will also enjoy the participation in a sort of 'virtual gym', a community-like rehabilitation environment possibly connecting fellow sufferers to increase inclusion and motivation.

Commercial gaming products like Nintendo Wii, Sony PlayStation Move or Microsoft Kinect exist that allow the user in a pervasive computing setting to act and interact with other users within a virtual environment, thanks to special interaction devices and suitable technologies to monitor the real environment and track the user(s) behaviour in it. These are proprietary solutions characterized by limited extensibility and flexibility. However, these gaming platforms have no specific features conceived and designed for a easy adoption in medical practice. Therefore, an ideal solution requires the following features and functionalities:

1. A set of exercises within a personalized, serious-games based, rehabilitation program, properly designed according to the patient's specific needs adapting dynamically and automatically to the user's behaviour/reactions. Ad hoc, real-time, configuration of the proposed physical and visual stimulations mediated by the virtual environment where the patient is acting.
2. Training of both patient and family members in the execution of the exercises. It must provide information for medical experts for off-line and on-line management of the rehabilitation protocol.
3. Acquisition and recording of relevant physical and physiological parameters, by means of suitable sensors providing a qualitative and quantitative picture of the patient's status and progress.
4. Promotion of social inclusion and community building by means of suitable Web 2.0 tools to prevent depression and stimulate patient's reintegration within daily life.

2.2   Mild Cognitive Impairments

There are several reasons why elderly people become reluctant to travel or even to take a walk. Besides physical impairment, orientation problems and the feeling of insecurity in terms of the fear of crime, often creates a lack of confidence in the ability to cope with 'easy' tasks that today are part of daily life. Examples of these daily activities are: - buying tickets for public transportation, taking money from the bank, going to the post office or even selecting products and paying for daily shopping.

To overcome physical impairments, innovative wheelchairs or other electromechanical aids are helpful. However, encouraging self-confidence in one's own abilities is necessary when providing home-based physical and cognitive training. Relevant information and location-based aids are also needed during outdoor activities.

Providing contextualized help in difficult situations reduces elderly people's reluctance when facing issues related to outdoor life. To achieve this requires different interoperable levels:

1. Home-based physical and cognitive training requires an advanced approach based on both innovative gaming products (e.g. Wii and Kinect) as well as on specifically developed Serious Games. This is to increase the engagement in the processes through fun and personal involvement.
2. Specifically conceived, usable and personalized orientation and navigation support has to address the decreasing ability of elderly people to orientate

in the surrounding environment as well as the fear of getting lost. Such a service will direct the users not only to different relevant places close to a current location but enable them to get back home as well. User behaviour and movements on different geographical scales need tracking features. For example, an intelligent doormat can help to prevent forgetting the latchkey when leaving the home. Indoor task-related reminders, public transportation access and usage can be supported by wearable and pervasive computing technology.

3. Improved social inclusion and accessibility through Web 2.0 technologies and tools have to create a user-friendly and highly interactive portal for social networks, chat, voice communication, professional aid and support for e.g. for buying theatre tickets.

The aim is to support elderly people during their daily life whilst encouraging their mobility both indoors and outdoors and giving users the feeling of not being alone whenever they need any kind of help, whilst giving them control on decisions and responsible choices. For this purpose again, we see Serious Games or exergames as an appropriate means for increased mobility as well as enhanced health and reduced social isolation. This requires dedicated services like a proper home-based physical and cognitive training and backend system to connect with caregivers or family members.

The technology needed is wearable and pervasive computing in a networked world providing protection of personal data whilst being connected with therapists, caregivers, relatives and friends. The operational infrastructure follows the abstract SOA (service-oriented architecture) model and the implementation uses self-standing components for the hardware platform and webservices and related technologies (WSDL, SOAP, UDDI) for the software architecture. This technological choice allows for high flexibility and reusability of software and hardware components. The operational infrastructure is the 'glue", allowing the proper and smooth functioning of relevant technical modules like a motivational and creative games engine, location and navigation tools based on GPS and GNSS, Web2.0 tools for social inclusion and communication, multimodal interaction like speech-based, touch screen, haptics, etc. The communication infrastructure uses. Wi-Fi, 3G, cabled, and Bluetooth etc....

The successful introduction of technological solutions strongly and critically relies upon meeting user needs. Nevertheless, the gap between the real help offered by proposed tools and the actual use of technological support can be wide. Therefore, the first relevant step when conceiving a new interactive system is to focus attention to human activities that can be supported by the technology and define the requirements to support such activities, thus applying a process that is known as 'human-centred design" [6]. This is an approach to interactive systems development that aims to make systems usable and useful by focusing on the users, their needs and requirements and by applying human factors/ergonomics and usability knowledge and techniques. The aim of such an approach is to enhance effectiveness and efficiency, improve human well-being, user satisfaction, accessibility and sustainability, and counteract possible adverse effects of use on human health, safety and performance.

It could be expected that an alignment with existing standards is sufficient (see [7], [8]). The fulfilment of the following criteria should thus apply to our target group:

- The design is based upon an explicit understanding of users, tasks and environments;
- Users are involved throughout design and development;
- The design is driven and refined by user-centred evaluation;
- The process is iterative;
- The design addresses the whole user experience;
- The design team includes multidisciplinary skills and perspectives.

However, in the course of our research[4] we found that the fulfilment of these usually applied criteria is not uniquely sufficient when dealing with the target group. Chapter seven in this book discusses this finding. To better understand the needs of the technology of wearable and pervasive computing and Serious Games we will briefly and at a high-level explain the concepts and insights we got from recent projects.

3   Objectives

In this book chapter, we briefly describe the remote healthcare computational architecture, Reference Rehabilitation Platform (RRP), which is based on distributed smart sensors (first processing layer) and personalized physiological assistive and intelligent algorithms integrated in a multilevel physiological human model.

---

[4] Joint AAL project ASSAM: http://assam.nmshost.de/; and Joint AAL project SafeMove: www.safemove-project.eu; both accessed 1.12.2015

We will describe the way the therapist can configure the Serious Games by using a web interface to set up the patient's individual care care plan. The set of games we chose for our research focused on the upper extremities; the approach itself is not limited to this.

We will furthermore report on experiences gained during the implementation and evaluation with patients in rehabilitation centres using different prototypes of the RRP implementation. For this evaluation, therapists and patients choose a set of mini games that therapists could adjust to the needs of the required individual patients' therapy. The games employed involved e.g. grasping virtual objects with a hand, or following a complex trajectory with the hands or the arm in an individually set area. These games required performing tasks accepted by the target group and configured to the patient's individual needs of physiological training and motivation. Subsequent chapters will describe the games in more detail.

4   Technology Description

The goal of defining a RRP is to provide a high-level technical view on the hardware and software infrastructure to support the development of Serious Games based rehabilitation. The RRP design follows the core principle of modularity whenever possible. We used a service-oriented architecture (SOA) with a set of core modules. Third parties can easily add additional modules (see figure 1).

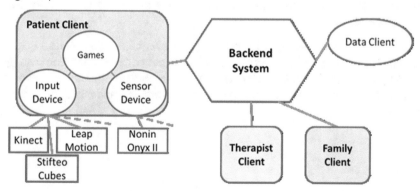

Figure 1 Reference Rehabilitation Platform (RRP)

We specified the functionality and data flow at an abstract level for the concrete technical implementation. The main building blocks of the RRP are the

Patient Client with the Serious Games implemented, the Family (caregiver) Client, the Therapist Client, and a Back-end system with the Central Information Space (CIS). The main interfaces are direct communication channels between the clients and data exchange channels between any client and the CIS. Thus, the CIS is responsible for storing and providing data for the overall system.

For the interaction, we use a 'Generic Data Interface (GDI)" which is the layer that abstracts from specific data formats and data schemes used by 'unforeseen" client games on the one hand and by 'unforeseen" external data providers on the other hand. The interface provides a generic way for sending and retrieving information from the CIS.

The *Therapist Client* developed using html and java scripts is a therapy management tool. It allows the continuous monitoring, management, and communication with patients during their rehabilitation activities (see figure 2).

The clinical staff use the Therapist Client to personalize and schedule the program exercises for each patient and for tuning the rehabilitation program on the fly. The purpose is to do long-term statistical analysis of collected data and information. This will additionally take into account the data produced by the 'Patient station' during patient's interaction with the games, such as motion data, games data and usage information. The analysis of the collected data supports therapists and clinicians in understanding the level of progress during the rehabilitation process and in choosing possible adjustments, both in the short and long term; thus, it provides a solid background for a more effective rehabilitation plan.

The Family Client, developed using html and java scripts, is a monitoring tool for family members or caregivers to support the patient's rehabilitation plan by over-seeingprogress and the most relevant information on therapy compliance of patients, as well as by communicating with therapists and clinicians in case of need.

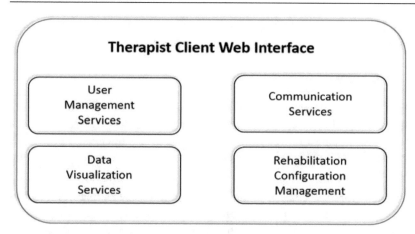

Figure 2 Services integrated into the Therapist Client

The *Data Client* component is part of the back-end structure of the Re-hab@Home platform. It retrieves data from the Central Information Storage (CIS) that stores the raw data received from the Game Component (GC) and the Signal Interpretation Component (SIC) for each game session. To highlight the problems during the gameplay, such as moments of high excitement etc., algorithms process the data. The challenge is to understand if the game intensity is e.g. too low or too high for the patient. To perform this kind of computation, we linked the Data Client to a relational database management system (RDBMS). It stores data useful for a data mining process designed to introduce a certain amount of 'intelligence" in data analysis. The quality and precision of the Data Client output are supposed to improve as soon as a sufficient amount of data coming from the game sessions is available for exploitation. We developed the Data Client as a command-line tool with the idea of being able to run on a server completely unattended: no interaction is required or in any way possible; it outputs a 'summary" stored in the CIS. The prime focus of its development was to keep the code at a high abstraction layer, not OS-dependent.

5   Developments

The developments covered the complete scenario with therapists, patients and caregivers for our user centred design approach [9] as well as the above-mentioned overall architecture and its implementation and evaluation. The implementation of the RRP we continuously adapted to the stakeholders' needs and evaluated it with patients and therapists in each development cycle.

5.1   Scenario

For a successful implementation, the involvement of all stakeholders is important: Thus, all technical specifications, design and development decisions and activities required the involvement of all the stakeholders who potentially might use the RRP for rehabilitation purposes: patients, therapists, families and caregivers.

We used workflow analysis and modelling as they play an important role in medical IT projects: Any implementation of an IT system requires an understanding of the processes involved and, depending on the scope and complexity of the system, will involve a certain amount of process redesign. We used meetings with the stakeholders for understanding the impact on the clinical work processes, defining the expectations and requirements for the IT solution, and managing the change process associated with the implementation.

The *Patient Station* is defined as the hardware and software installed at the patient's home for rehabilitation exercises thanks to special interaction devices and technologies. During the course of an exercise performed by the patient, sensors provide extensive streams of measurements. All the games a user can select are set with their specific parameters by a therapist, according to the patient's engagement and rehabilitation condition. Furthermore, in parallel with all the rehabilitation exercises, presented as games, the patient can communicate both with the therapist and relatives (or caregivers), through a specific 'messages' function.

5.2   Implementation

There was the need to study different COTS devices for understanding their functionalities and compliance with regard to the target patients. For instance, the Wii controller is very difficult to handle as patients often have problems grabbing objects. Also correctly tracking the movements from these kinds of patients with the KINECT V1 was hard.

We tested the LEAP Motion for hand and lower arm tracking. Tracking the vertical movement with the Kinect V1 was not sufficient for the purpose. The LEAP Motion has strong potential for rehabilitating fine movements of the hand and also for the arm. The Therapist Client assists the professional user over the different phases of the therapy, both remotely and locally. The therapist creates the care plan for each patient, selecting the games and all the parameters for each of these games. This requires an easy to use configuration for proper user acceptance. While patients perform their rehabilitation plan over the time, the therapist can access data collected during therapy for evaluating the possible progress and fine-tuning the plan with the kind of games, parameters, etc. The therapist can communicate with patients and caregivers using the 'messages' function too. The caregivers get an overview of the possible progress and the therapy compliance of patients. They can communicate with the patient and the therapist.

Different input devices were available to use SDK or direct drivers connecting to the *Patient Client*. In the end, trying to develop a standard component suitable for each device proved not to be a smart choice as it forced us to ignore specific features provided by a individual devices. Choosing to exploit specific drivers for the devices led to simpler game design, minor overheads in communication and maximized the feature availability in each game.

Technology evolves at a really fast pace. Therefore the choice was not to develop very low level data stream elaboration but to exploit evolving technologies like Microsoft Kinect V2 sensors.

5.3   Evaluation

Each year we finished one design cycle ready for evaluation. We did a technical validation of the basic pilot in the first year. We assessed the usability and graphical user interface (GUI) for the enhanced pilot in the second year. We valued the final demonstrator at the two clinical partners of the project in trials with twenty patients in twelve subsequent sessions over a couple of weeks. We further evaluated the system with six therapists and caregivers.

In the first evaluation the devices and games proposed to patients, caregivers and therapists were already positively accepted. Several indications for improvements were provided, mainly fostering the need for supporting calibration of the solutions to the range of motion and specific needs of patients, customization of the games, visual and audio elements of the games, motivational strategies to better engage patients and provide feedback on progress during therapy, and support for collaborative forms of play. In each iteration during the evaluations of the clinical interface for therapists, advancements based on concept and usability were uncovered.

At the end of the development, we achieved a working system ready for evaluation in clinical sites (see book chapter eight). The system is also ready for other sites than the ones of the clinical partners of the project and for exploitation outside the predefined therapeutic area of the upper extremities. Moreover, we collected data as a starting point for later use in clinical studies evaluating other or extended therapeutic approaches. Specific issues of the target group deal with depression episodes, cognitive dizzy spells or slight limitations; see book chapter seven on experiences with that user group. The success of any therapy strongly depends on the emotional state of the patient. Knowledge concerning the emotional state is thus very important; see book chapter nine on methods to determine the emotional state.

6   Results

The results we achieved are based on three pilots as the outcome of three development cycles following the user centered design approach.

6.1   First Pilot

The aim of the first pilot was to test a first set of game-based solutions for in-home rehabilitation of patients within a simple scenario and a basic technological platform. This was important for an understanding of which kinds of games and input-output interfaces could be valuable in a therapeutic context, suitable to home environments, targeting patients with low disability or difficulties with Activities of Daily Living (ADL) arising from the upper body. This initial testing phase has allowed us to assess the usability and ergonomic qualities of the solutions considered, to further elicit and understand patients' needs and rehabilitation constraints, to refine these initial solutions and develop a more comprehensive set of functionalities for the next pilot phase.

A small sample of six patients (4 post-stroke, 1 MS and 1 motor impaired) with motor impairments of the upper body were involved within individual sessions of up to two hours of duration. The interactive sessions with the games tested were video recorded for subsequent analysis; patients experienced the different input devices and games in a random order as they arose during the ongoing session. Whenever possible, a therapist and a family member attended the session to provide feedback on the games from a therapeutic and caregiver view, respectively.

In total, we evaluated eight games and received valuable data concerning the scope of the games as well as the needs for configuration. Out of this evaluation of the *Patient Client*, we got the following requirements:

1. Enable games to be easily adapted to patients' different needs and range of movement;
2. Ensure correct execution of the movements, detect compensation;
3. Engage patients by providing more variability and difficulty levels in the games;
4. Check audio and visual elements for accessibility and usability;
5. Add motivational elements and incentives;
6. Add collaborative forms of play.

Three therapists conducted collaborative walkthroughs of some initial mock-ups of the *Therapist Client* using wireframes exploring the usability, look-and-feel, and comprehensiveness of the features prototyped. The main requirement was a more intuitive and design focused on essentials. The therapists further required visualizing the progress achieved by the patient in each session and extending the design of tele monitoring and communication features with patient and caregivers in order to facilitate in-home therapy and information exchange.

6.2   Second Pilot

The aim of the second pilot was to test a refined and extended set of game-based solutions for in-home rehabilitation of patients with motor impairments of the upper body. We assessed whether usability and motivation related problems improved compared to issues identified in the game solutions evaluated during the first pilot. Furthermore, we aimed to support the final integration of the games and clients of the RRP in preparation of their deployment and evaluation in the third year trials.

Three patients (age range from 58years old to 81years old) with motor impairments of the upper body (1 post-stroke, 1 MS and 1 Parkinson Disease) were involved in sessions with the same overall setting as of the first pilot.

Here are the findings (see also Table 1): All three patients could play at least one match with each device-game combination proposed. They maintained their interest and engagement with the games experienced over the complete pilot session. All patients experienced some problems in interacting with the Kinect V1 device to select the next game to play through the Patient Station menu. Two patients had difficulties properly controlling the LEAP Motion device. Two patients had some problems in identifying the colours/elements displayed on the Sifteo Cubes, due to the low resolution of the cubes display and their visual impairments. However, all patients were, supported by the pilot moderator, able to understand the games mechanics and instructions with individual differences in the preferences for the games.

Table 1 Mean value of patients replies to the Quality of Instructions/communication, User Experience, Motivation questionnaires on Likert scale 1-5 (5=most positive level).

Patient	Quality of Instructions	User Experience	Motivation
Patient 1	4.33	3.5	3.5
Patient 2	5	4.5	5
Patient 3	4.33	4.12	4.25

Overall, patients preferred games using the Kinect and LEAP Motion. The patients did not favour the Sifteo Cubes. Patients perceived the Kinect and LEAP Motion games as being more useful and relevant to motor rehabilitation, while they considered the Sifteo Cubes games more useful and relevant to cognitive rehabilitation. Five therapists with one to six years of work experience in the rehabilitation field conducted collaborative walkthroughs of the working prototype of the *Therapist Client*. They were familiar with gaming products like Wii or Kinect. The therapists gave, compared to the first pilot, more detailed suggestions for improvement of the following features:

1. Exercise assignment configuration;
2. Communication with patients;
3. Calendar with more than one session per day;
4. Games and equipment catalogue;
5. Visualization options for monitoring rehabilitation programs of patients based on different examples.

This outcome became part of the specification of the third pilot developed.

6.3   Third Pilot

The two clinical partners evaluated the third pilot with ten patients at each site and within twelve sessions per patient. A database containing the data of the trials of the final prototype was set up for statistical analysis by the medical partners. The tests proved the high usability of the RRP. The implemented games succeeded in keeping the motivation up for the patients over the full duration of the study; i.e. twelve sessions of two hours each within two to three weeks.

The framework for the evaluation of the user's physical progress and preserving is the bio-psycho-social model of the International Classification of Functioning, Disability and Health (ICF) [10]. The use of the ICF framework ensures a holistic approach to functional evaluation of functioning and the necessary functional changes as well as eventual changes in functioning in response to the RRP based rehabilitation.

One measures the progress according to the ICF at three levels: at the level of body function, level of activity and level of participation. The medical and technical partners involved published the outcome of their clinical research using the RRP separately, as this would go far beyond the scope of this book chapter. However, book chapter eight provides an overview of these findings.

The plan is to use the database set up for the final trials no longer than two years after the end of the project (2016/17) with the purpose of of not only monitoring the success of the individual therapies but also to collect data for statistical analysis.

Nevertheless, the RRP, by the end of the project, had already proved to be applicable in such an interdisciplinary setting and can fulfil the diverse requirements from therapists, computer scientists and medical scientists. We could adapt the system within three years to different needs without any redesign of the overall RRP architecture. Moreover, the RRP can be used for other therapies with Serious Games and COTS sensors as input devices of either motion, speech or any physiological data without changing the overall approach.

7 Business Benefits

There are constant rapid advances in modern medicine. New treatments and devices appear frequently. We have to be careful in the healthcare system to evaluate each of these new solutions based on a wide set of criteria. Only in this way, one can reach a decision whether or not to finance and implement it.

In European countries, we have great differences in healthcare policies and funding schemes. In Austria and for most people in Germany, the institution that might be willing to pay for rehabilitation is the national social insurance system. The citizens receive state of the art medical treatment (which includes rehabilitation) as part of this insurance free of charge. Only in very rare and convincing cases, the patient is willing to contribute financially to receive 'better' services.

The healthcare system decides on the cost-effectiveness of all competing alternatives. The main goal of cost-effectiveness analysis is to test the hypothesis that the mean cost of one health care intervention per defined utility unit is smaller than one or more competing alternatives requiring a very precise definition of the utility unit, making sure the analysis is indeed reflective and accurate. This makes a general approach within a research project more than difficult. Patients could use the RRP in two-hour sessions in a rehabilitation institute or at home. One has to compare extended hospital admission with indicators such as mean duration, employee costs, overheads etc. and home rehabilitation with indicators such as nurse visits, system's maintenance cost etc.. Stationary rehabilitation, due to the national insurance systems of Austria and Germany, is usually limited to two to three weeks. We observe that rehabilitation after a stroke usually takes months or even years of continuous effort. Thus, our approach is new; we cannot compare it to the usual stationary approach with a simple cost-effectiveness analysis.

We therefore developed a detailed model with key performance indicators used for user's physical progress and sustained status allocated for clinical indicators of the RRP effectiveness as opposed to hospital admission. One can use these indicators as an effectiveness measurement in order to define an effectiveness unit.

Home rehabilitation systems are an growing trend in the healthcare industry, although still somewhat limited. They lack the attractiveness compared to games for entertainment only; there is always the risk of becoming boring to the patient after a short period. They require technical or medical expert knowledge when being set up. They provide only limited ability to adapt to the specific patient needs and target one specific type of rehabilitation only like stroke, cognitive stimulation, etc. Furthermore, there is a lack of details on possible costs and have portability issues involved with deployment in real life settings.

In our post session interviews of stakeholders and patients, we also checked purchase intentions and price ranges. With health care regulators, insurance companies and retailers we determined whether they provide support to cover or subsidize costs based on value for money (see [11] for details). We created six unique sets of interacting key performance indicators (KPIs) reflecting aspects like user related, technological, medical, organizational, environmental and financial.

We added value for the end users and stakeholders to gain a competitive business advantage and took the chance that the RRP will also be used by others on a globally large-scale distribution for a positive impact on the entire rehabilitation sector.

8 Conclusion

The Reference Rehabilitation Platform (RRP) for Serious Games described in this book chapter is suitable for the development process as well as providing a working prototype on-demand for evaluation, marketing or even business talks on further exploitation.

Therapists, patients and caregivers can, with the help of the RRP configuration, use and monitor the rehabilitation process using Serious Games in the patient's home. Although designed for the rehabilitation of the upper extremities, the RRP allows the integration of any sensor able to monitor motions of the whole body, the hands and fingers or even speech. We developed various indicators and procedures for evaluation and assessment by the different stakeholders. The leading concept was the user-centred design approach (UCD) for user-friendly products or systems designed to avoid potential resistance or reluctance from using it when available. The outcome is an open access system for further use and a database to collect data for the evaluation of rehabilitation therapies using Serious Games.

One requirement is that the underlying hardware is commercial off-the-shelf. Although the Reference Rehabilitation Platform (RRP) presented here is universal, it requires the integration of actual hardware devices that are not yet available for all kinds of rehabilitation purposes. In the next chapter we will show what was available in the market at the time of the project and how we approached the possible solution for evaluation.

9 References

[1]   Clark C. ABT: Serious Games; University Press of America, ISBN 978-0-8191-6147-5, 1987

[2]   Rego, P. Moreira, and L. Reis: Serious games for rehabilitation: A survey and a classification towards a taxonomy, in Information Systems and Technologies (CISTI), 2010 5th Iberian Conference on, 2010, pp. 1 –6.

[3]   EVERS, Silvia MAA, et al. International comparison of stroke cost studies. Stroke, 2004, 35. Jg., Nr. 5, pp. 1209-1215

[4]   TRUELSEN, T.; EKMAN, M.; BOYSEN, G. Cost of stroke in Europe. European journal of neurology, 2005, 12. Jg., Nr. s1, pp. 78-84.

[5]   WALLACE, A. C., et al. Standardizing the intensity of upper limb treatment in rehabilitation medicine. Clinical rehabilitation, 2010.

[6]   MAGUIRE, Martin. Methods to support human-centred design. International journal of human-computer studies, 2001, 55. Jg., Nr. 4, pp. 587-634.

[7]   ISO 9241-210:2010 'Ergonomics of human-system interaction - Part 210: Human-centred design for interactive systems'

[8]   ISO 13407:1999 'Human-centred design processes for interactive systems"

[9]   Gulliksen, Jan, et al. 'Key principles for user-centred systems design.' Behaviour and Information Technology 22.6 (2003): 397-409.

[10]   World Health Organization, et al. Atlas: child and adolescent mental health resources: global concerns, implications for the future. World Health Organization, 2005.

[11]   Michael Lawo, Peter Knackfuß, Zvika Popper (2015): Measuring organizational acceptance of a computer game based system for home rehabilitation; Research Gate, DOI: 10.13140/RG.2.1.1032.6249.

Acknowledgment

The authors of this chapter thank all their co-workers and funding agencies of the many projects in recent years. Special thanks go to Dominik Angerer of the project partner Netural in Linz/Austria, Gabriel Benderski of the project partner Edna Pasher in Tel Aviv/Israel, Silvana Dellepiane of the project partner University of Genua/Italy, Silvia Gabrielli of the partner Create-Net in Trento/Italy, and Lucia Pannese of the partner Imaginary in Milan/Italy and the former doctoral students Hendrik Iben and Ali M. Khan of the second author; without the support, this chapter could not exist.

III Hardware and Software for Solving the Serious Game Challenge
Using Commercial Off-the-Shelf Components and Open Source Software Frameworks

Hendrik Iben / Ali Mehmood Khan / Michael Lawo

Abstract

The purpose of this book chapter is to show how to solve the problem of selection of an appropriate hardware and software. This is a challenge for any non-standardized application domain and a problem any research project has when looking for a general purpose solution for a specific problem. Here we target the evaluation of a Reference Rehabilitation Platform (RRP) for Serious Games. One constraint in such a case is that all components should be commercial off-the-shelf during the runtime of the project. Components should be well tested, provide sufficient firmware and documentation for integration and have the potential of becoming a kind of standard in the market. This is essential, as the focus of any such project is the problem solution and its evaluation of the hardware and software. We propose a platform where different Serious Games can be deployed and input devices as bio sensors can be plugged in easily. We developed a solution where these components can be replaced by new components easily without changing the whole architecture. This book chapter addresses technical issues and provides an idea how to integrate Serious Games for rehabilitation purposes as described in the chapter four.

© Springer Fachmedien Wiesbaden GmbH, part of Springer Nature 2018
M. Lawo und P. Knackfuß (Hrsg.), *Clinical Rehabilitation Experience
Utilizing Serious Games*, Advanced Studies Mobile Research Center
Bremen, https://doi.org/10.1007/978-3-658-21957-4_3

1   Introduction

The benefits of rehabilitation can be expressed as a higher quality of life for patients as well as their families. Additionally, rehabilitation can result in lower costs for additional health care and higher productivity, as patients may return to their normal lives sooner. Moreover, healthcare innovations that improve rehabilitation could increase the benefits even further.

One of the latest innovations in rehabilitation is the use of Serious Games for cognitive, psychological, motoric, and social rehabilitation. Rehabilitation gaming is similar to tele-rehabilitation which is mediated by videophone [1] and rehabilitation mediated by Virtual Reality (VR) ([2] and [3]). The main benefits of mediated rehabilitation in comparison to traditional rehabilitation, particularly game-based rehabilitation, are associated with the motivation to engage in rehabilitation, the objectivity of rehabilitation measurements and the personalization of the treatment.

2   Challenges

When designing technical solutions within research projects, beside technical issues, issues of the usability of any proposed solution are predominant. However, when the solution targets a rehabilitation issue, in addition to technical knowledge, medical and therapy knowledge is also required. In the context of rehabilitation addressing social aspects is further required as in addition to the patient, other stakeholders like medical doctors, care givers, family members are involved. In such an environment, the technical solution must be there but should not cause extra problems. Solving the research problem is in the main focus but technical issues are also always there. Here, it helps to build on pre-existing knowledge within the research team of existing technical solutions and experiences with technology, devices and tools needed.

When using Serious Games in rehabilitation settings, developers should keep in mind that patients do not want to wear devices all the time. However, in some cases patients have to wear a wrist band so that the system can recognize an emergency case. It is always hard to integrate new components e.g. a new game or a new input device to an existing solution. Requirements are not static; they might be changed in the future due to results from a preliminary evaluation. Sometimes new games are required as a new therapeutic target needs to be addressed. New input devices might come to the market, which may lead to discarding already developed serious games and to developing a new system from scratch.

We know about those challenges from being involved in many research projects over recent years. Thus, developing a platform for Rehab@Home was not a completely new challenge. From a technical side, the platform has to provide the capability to add more games, input devices and bio sensors, allowing the researches to be flexible and to answer the requirements defined by the different stakeholders. An architecture fulfilling this requirement will provide developers of the research problem solution an environment where they can focus on their issue, e.g. to develop a new game, use data for a further evaluation or test another interaction device or sensor to monitor medical data.

In our project, we investigated the issues related to long-term physical / cognitive rehabilitation processes to identify suitable technical solutions which might enable elderly people to enjoy high quality rehabilitation for a much longer period than the health system can currently afford.

Using standard hardware components and devices was a main challenge alongside many others like suitable medical data processing algorithms, personalized and serious-games based rehabilitation pathways or Web2.0 social and communication tools. The basic approach is inspired by existing commercial platforms, like Wii Remote and Kinect that allow the user to act within a virtual environment and interact with other users.

The hardware and software approach seeks to provide exercises and training based on serious-games within a personalized, user friendly and engaging rehabilitation program. Constraints are based on the need to be cost effective without a bulky infrastructure and integrated sensors able to collect relevant physical and medical parameters. As the system in mind includes patients' status inspection and relapse prevention it has to support off-line and on-line management and monitoring of the rehabilitation protocol and promote patient's social participation and community building. The goal is to make the patient's home the place where physical and cognitive rehabilitation process can be performed in an intensive and engaging, though properly controlled way, whilst promoting social inclusion and quality of life.

The four components of the Reference Rehabilitation Platform as described in the previous book chapter are the patient system, the therapist client, the family client and the back-end system (see figure 1). The patient client is for patients where therapists will deploy the games and patients need to play the suggested games whilst wearing physiological devices. The therapist client will be used by therapists or medical experts where they will monitor patients' activities and health conditions. The family client will be used by patients' family members. They will be able to see limited information. The back-end system is meant for storing patient history including physiological data, activities, game scores etc. Only the patient client needs to interface different sensors for interaction within the therapeutic Serious Games as well as for the monitoring of physiological data. A suitable approach for this patient client in the context of the Reference Rehabilitation Platform is described in this book chapter.

As these systems are developed by computer experts but will be used by doctors, patients and patients' relatives, these users are unlikely to have any idea about technology. Therefore, training is essential for all users and stakeholders.

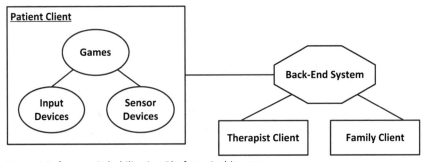

Figure 1 Reference Rehabilitation Platform Architecture

3 State of the Art for a Technical Approach

As mentioned in the first chapter of this book, the combination of game-based concepts with ICT has application domains besides military training [4] education [5] and healthcare ([6] and [7]). Serious Games [8] mean the use of computer games beyond the purpose of entertainment to achieve certain high-level goals rather than for pure fun. The entertainment mechanisms are used to stimulate the user's engagement since the traditional treatment approaches are typically repetitive and not appealing for patients. Therapists expect to overcome low compliance to the prescribed therapy sessions and frequent drop-outs. The potential for more precise assessments of the patients' performance in comparison to human observers is a further benefit for medical doctors and therapists.

The Serious Games concepts used for rehabilitation purposes differ in a number of aspects from the kind of rehabilitation (such as disease, injury, or mental skills) and skills targeted to the interaction designs and technologies used. Some commercially available rehabilitation systems exist that focus on a specific set of actions using one chosen sensor technology. These systems tend to be closed to third parties and are not meant to be extensible ([9], [10], [11], [12], [13], [14], [15], [16] and [18]).

The RRP aims to provide an open platform for solving different problems. This is enabled by a framework for integrating suitable devices into the system. It implements a general mechanism for transmitting arbitrary inputs from medical devices like interaction sensors.

Generic approaches for using sensor data in a larger software system have been subject to scientific work but are not yet available as a common part of a software environment. Prominent examples of these systems are the Context Recognition Network (CRN) Toolbox[1] by the University of Passau and the ETH Zürich and the Context Toolkit (CTK) by Anind Dey[2]. The CRN Toolbox provides predefined components for accessing various sensor devices and flexible ways of processing the acquired information while the CTK provides a framework for accessing sensor data over a network featuring a search for sensors by description. However, both tools have not been actively maintained for several years.

A more recent, but also less feature-rich approach, is found in the national funded SiWear project[3] where simple sensor information is made available to mobile clients in real time. The TZI Context Framework (TCF)[4] developed by the first author of this book chapter at the TZI of Bremen University [19], provides network access to sensor information much like the CTK but approaches the problem by using a simple client-server concept instead of creating services out of each sensor device. It uses a custom communication protocol on top of TCP/IP to minimize the overhead while transmitting sensor data.

The technical responsibility of the platform is to provide an environment in which other components can be easily integrated to work together. This is achieved by defining a common set of technical agreements that allow developers to focus on their specific problem such as acquiring sensor data, serving

---

[1] http://crnt.sourceforge.net/CRN_Toolbox/Home.html, last accessed 10.8.2017
[2] http://contexttoolkit.sourceforge.net/, last accessed 10.8.2017
[3] http://www.siwear.de/, last accessed 10.8.2017
[4] https://github.com/wearlab-uni-bremen, last accessed 10.8.2017

medical data to therapists or managing the flow of a serious game without needing to coordinate with other components.

As a general approach to this goal, the paradigm of service oriented architectures (SOA) can be used. The idea behind this approach is to provide a technical framework – a container – that is able to host and make available services to other components. These services provide the necessary functionality for the system and define interfaces in a common way for the framework. If a component needs a specific functionality, it accesses the corresponding service via the container using the defined interface.

This organisation has the benefit of making services interchangeable and abstract from the implementation of the service. Furthermore, the container concept allows dynamic distribution of services, making the system flexible and maintainable.

A suitable container platform is offered by Apache Tomcat. Whilst many types of this project exist, the core of Tomcat is a Java based server that manages Web-Services based on Java EE implementations. The use of Tomcat overcomes the problem of the increased start up time of Java applications (compared to native applications) by keeping the services in an active state. Tomcat also follows the Inversion of Control (IoC) paradigm allowing a very easy configuration of many aspects of the container itself and also of the hosted services.

Tomcat itself contains a HTTP server and can therefore be used to create pure web based applications but it can also be used in conjunction with a separate server if needed. Services do not need to provide content suitable for a web browser but can make use of the very common HTTP protocol for communication, e.g. as a RESTful service, via SOAP or by providing JSON data.

As a technical contrast to this, the platform will also have to facilitate the transport of real-time sensor data from the environment to various software components. Whilst providing this information as a service is conceptually a promising approach, the technical protocol used for web services is not designed for transmitting this kind of information and would result in large overheads in data volume.

The problem of distributing sensor information in a SOA-like environment has only been addressed in a few works. The specialized client-server approach enables low latency and low overhead transmission of sensor information. This approach had promising results in pick-and-place applications where information from environment sensors was used to drive a mobile workflow process. This approach can be treated as a special service case that requires a special protocol but is only needed by a few components.

The largest benefit of SOA is the abstraction of functionality into services and the provision of a common interface to access them. Whilst this allows an existing system to be extended later, it also means that a common set of services has to be always present and known to each component as a starting point. In the most generic SOA this could be reduced to a single service that provides the functionality to search for other services and having each component register itself with this service to make it available for others to use. This approach is however not very practical when many aspects can already be defined as needed for all the intended application scenarios.

As mentioned before, the Tomcat container offers an IoC mechanism that allows users to configure services at runtime. This mechanism can be used to identify the location of necessary services while the communication follows the abstract specification defined by the functionality of the service.

For some aspects, a service registry might be an interesting concept. However, as existing registry components are generic and have been developed with globally interacting companies in mind, they tend to become very complex to use. Therefore, one should consider implementing simpler and more practical solutions to locate specific services if the need arises.

4 Interaction Devices

As mentioned above, one of the requirements for our research approach was the use of commercial off-the-shelf components. This was additionally applied to the active user interaction devices we took into account e.g. the Kinect sensor, the Wii Remote and Leap motion, as well as for the passive user interaction when using the e-Health sensor platform. These components were not specifically designed for the application domain of rehabilitation and are far from being standard components. However they do fulfil the requirement of being commercial off-the-shelf. In addition, a computer with small form factor for data processing and a flat screen as an output device are standardized general purpose devices.

4.1   Kinect Sensor

The Kinect sensor is a gesture and posture based input device for the Xbox gaming system manufactured by Microsoft (see figure 2). It features a video camera combined with a depth camera that provides distance information for each point in the video image by sending out structured light in the infrared spectrum and interpreting the resulting pattern. The Kinect is interfaced using the USB standard and a proprietary data protocol. It can be accessed without the gaming system by either a SDK provided by Microsoft or third party software. Besides the image information, the Kinect also has a built-in microphone that can be used for speech commands. The main feature of the Kinect Sensor, combined with the software on the computer, is the construction of a 'skeleton' of the user. A simplified model of the human body is constructed from the depth image that provides the software with the positions and orientations of all limbs. During the project runtime, Kinect Version 1 was replaced by Version 2 with improved features and easily integrated into the overall architecture.[5]

Figure 2 Kinect Sensor

4.2   Wii Remote

The Wii Remote, commonly known as the 'Wiimote', is the standard controller used by the Wii gaming system manufactured by Nintendo (see figure 3). The Wiimote features a three-axial acceleration sensor, an IR tracking system, a vibration motor, a speaker and various buttons. It uses a standard Bluetooth connection for communication and can be easily interfaced using this standard without the need for the complete Wii system. For data transmission, an L2CAP link[6] is established to the Wiimote and data packets are sent to and received

---

[5] http://msdn.microsoft.com/en-us/library/hh438998.aspx

[6] https://www.bluetooth.com/specifications/assigned-numbers/logical-link-control

from the device. After an initial setup, the device streams sensor data to the host at a variable rate.

There is a maximum of 100 data packets per second each containing all three axes of acceleration and lower rates are observed if no change in acceleration is detected. Some Bluetooth implementations, especially for Java, might require setup as the Wiimote uses reserved port numbers with Protocol Service Multiplexers (PSM) like a common Human Interface Device (HID) device but does not adhere to the standard. Therefore the Bluetooth stack on the computer side might refuse to connect to the device without special preparation. For infrared (IR) tracking, the Wiimote expects two IR light sources on top or below the display. This light source is called the 'Sensor Bar' but is really just a collection of LEDs.

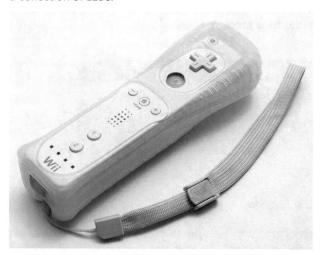

Figure 3 Wii Remote

The information from the IR tracking can be used to estimate the distance to the screen and to steer a pointer. To use the vibration motor and the speaker, special commands are sent to the Wiimote. However, playing audio is problematic when not done by the Wii console and success varies. This limitation is due to the lack of documentation from Nintendo and probably also as a result of non-standard features of the Bluetooth chipset in the Wii system.

Add-on-modules for the Wiimote exist such as the MotionPlus (a three-axial gyroscope), the Nunchuck (a joystick and a second three-axial accelerometer for the other hand) and the Balance Board (a scale-like device that measures weight and the users balance) that offer additional information for motion input. Newer Wiimote controls have the MotionPlus already built in due to its broad use in the Wii gaming industry. Data from these modules is either streamed along with the normal Wiimote data using a defined position in the data packets or interleaved within the stream of packets.

4.3   Leap Motion

The Leap motion device was a relatively new input device when starting the project. It is based on the recognition of gestures and intended to be placed in front of a computer monitor to track the hands of the user. The idea is to extract a detailed model of the hands, including the individual digits, and analyse the motion for predefined gestures to execute actions (see figure 4). More information can be found at the product website.[7]

Figure 4 Leap Motion

4.4   E-Health Sensor Platform

With the e-Health Sensor platform (see figure 5), many medical devices are made available for experimentation. The platform itself is not self-sufficient and needs a suitable component to process the included sensors. The available documentation allowed the platform to be used as a commercial off-the-shelf product, although not as a medical device.[8]

---

[7] https://www.leapmotion.com
[8] http://www.cooking-hacks.com/documentation/tutorials/ehealth-biometric-sensor-platform-arduino-raspberry-pi-medical accessed 10.8.2017

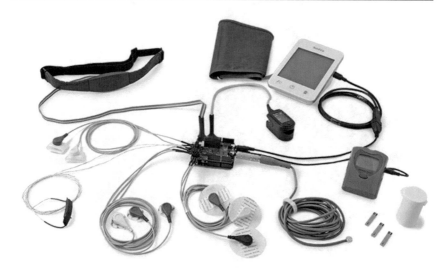

Figure 5 e-Health Platform

4.5   Main Platform

Because of its small form factor and experiences from previous research projects as a computer system, the Zotac ZBOX ID89 Plus (see figure 6) was used as commercial off-the-shelf device fulfilling all requirements of a standardised product.[9] This product was chosen as having enough computational power to allow for experimentation with various devices. It features an Intel Core i5 dual-core CPU running at a maximum frequency of 3.6 GHz. For connectivity it provides USB ports, Bluetooth and Wireless LAN that allows us to connect various devices without additional hardware. It can be connected to a standard PC monitor or directly to a TV screen via HDMI connectors. We choose Windows 7 as the operating system as the current prototype software needs a Microsoft Windows operating system. Within the evaluations with the Kinect and Leap Motion, we used the USB ports of the computer.

---

[9]   http://www.zotac.com/de/products/mini-pcs/zbox/intel/product/intel/detail/zbox-id89-plus-1.html accessed 10.8.17

Figure 6 ZBOX ID89 Plus

4.6   Raspberry Pi

The Raspberry Pi (figure 7) was chosen as a platform to process sensor information to be transmitted to the main platform. This computer has a small form factor and consists of a single circuit board. It features an ARM CPU and various connectors for interfacing devices. In contrast to a PC platform, it was designed to interface various hardware on a low level and provides the needed general purpose IO (GPIO) pins.

RASPBERRY PI MODEL B

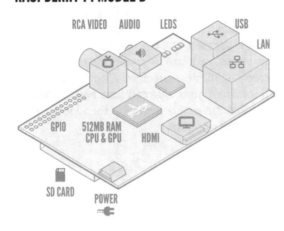

Figure 7 Raspberry Pi Diagram

Whilst the device is quite powerful and has been used in different previous research projects, we could not use it as platform for the Serious Games application here, as it has a Linux operating system and does not support Microsoft Windows or Mac OS systems.[10]

5   Methodology

The test platform provides an infrastructure to deploy games running on the platform. The deployment of games is not meant to be done by the user of the system but at the time of setting up the system for an individual patient by a therapist trained for this purpose. Each game acts as a service to provide a common way of accessing it, although the platform expects the game to take control over the machine when the corresponding interface is used.

One of the core components of the platform is therefore the Game Registry that keeps track of available games and their dependencies. This component provides a graphical user interface to the patient that serves as an entry point to starting the appropriate games.

Since the games make use of sensor data, there is also a Sensor Registry that can be queried by other components for information on available data sources. Sensor data is generally not transmitted via a web-service interface but via a separate server component. The Sensor Registry provides components with information on how to access specific sensors.

A third component is the server that transmits sensor information. This specialized server deals with distributing real time sensor data to interested components.

All components that are exposed as services are hosted in the Apache Tomcat service container.

5.1   Extendibility

Since advances in technology continuously lead to better suited devices for rehabilitation at home, it is necessary to keep systems open for new components. The most likely scenario of an extension is the introduction of a new serious game to offer a new form of rehabilitation. If such a new game can

---

[10] http://www.raspberrypi.org

merely reuse the previous sensors and provided information, this extension is simply realized by adding the new game component as it will find the services it needs whilst making use of the established interfaces.

A more complex scenario arises when a new kind of sensor or other device is to be added to the system. This can either happen along with a new rehabilitation game or to replace a previous device. In both cases, the new device will be similarly added to other devices, thus enabling future components also to make use of it. Since for most sensors a web-service representation is not desirable, sensors will be relayed over the context server component and the presence of the sensor will be registered.

A third common example is the integration or replacement of software modules. After integration into the container, only minimal configuration is necessary to make other components use the module. This is possible due to the anticipated inversion of control scheme used in the platform.

As with the core components, actual technical needs of extensions are unknown beforehand. But the Service Oriented Architecture (SOA) is based on Java Enterprise Edition (EE) technology and while commercially solutions supporting every feature of Java EE exist, we concentrated on the usability of open source solutions that can be easily tailored to needs.

A proven service container for Java EE is the Apache Tomcat server[11] that only supports a limited subset of features but can be extended with additional libraries. To create web services that communicate via Simple Object Access Protocol (SOAP) and Web Services Description Language (WSDL), the Apache Axis library is added. Further components exist that provide object persistence, database access and many other features that can be integrated when needed.

Since many tools exist to help in rapidly creating web services using the Tomcat container, and the ability to only integrate necessary features to save resources on limited devices, this technology made the ideal choice for our project. Also, the additional benefit of being a cross platform technology did not enforce a specific underlying system onto the rest of the overall system.

---

[11] http://tomcat.apache.org/

5.2 Initial Framework Prototype

With the start of initial integration efforts, where components from our technology inventory were combined into prototypes, a flexible software platform was needed. The first demonstration concepts had to deal with integrating information from sensor devices with serious game software. As a step between an ad-hoc implementation and a fully configured web-application container solution, we decided to implement a flexible base application. This application used the Spring framework[12] as a configuration mechanism that is also found in web containers and can therefore be developed further to be integrated into a complete system.

The responsibility of the initial framework was to provide a defined way of adding components to the system. Like a web application container, new components were put into the framework and set up by configuration files. This allowed for quickly changing components without needing software changes. We choose the Spring framework as it provides all the necessary functionality for this. All we needed to add was a very simple application to start and perform some convenience tasks for starting and stopping services.

6 First Prototype

For our initial integration, a game-like application created in the Unity Game Engine[13] is designed to make use of information from a Wii Remote Controller (see figure 3) and sensor readings from the e-Health platform (see figure 5).

The application is a very simple implementation of the classic *Pong* game where the player controls a paddle at the bottom of the screen and has to keep a bouncing ball from touching the bottom. Additionally, but without function, the heart rate information from the player is displayed (see figure 8).

The game does not keep a score and cannot be won. It served as a technical demonstration for using external input.

---

[12] https://spring.io/

[13] http://unity3d.com/

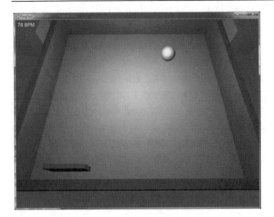

Figure 8 Pong game

6.1 System Architecture

Games made with the Unity Game Engine can contain scripts in various programming languages that can be used to receive non-standard information. For
the demonstration, the game used this mechanism to receive information for
controlling the paddle and reading the heart rate sensor over a network connection. This allows external components to control the game without needing
to know any details of it and also the game does not need special preparation
to use sensor devices.

For encoding information, we choose the popular JSON[14] format that can be
used to transfer arbitrary data and is available for the majority of programming
languages used. The game only reacts to two pieces of information from a received JSON message, the value of a 'hPos' attribute that defines the horizontal
position of the paddle and a 'bpm' attribute that defines the current heart rate
of the player. As a convention, the horizontal position is encoded as a floating
point value, where -1 represents the most left, 1 the most right and 0 the centre position. Heart rate is interpreted as a decimal number.

The game itself cannot be integrated into the platform (see figure 9) as the
Unity Engine has no concept of being started externally. The abstraction via
network communication however mitigates this problem as the platform only
has to provide information but does not need to check if it is actually used.

---

[14] http://json.org

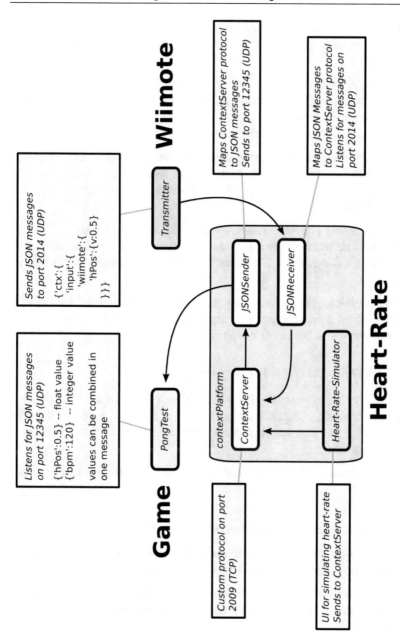

Figure 9 System Overview

The system consists of a central server that receives sensor information and a number of clients that either send this information or receive updates on sensor changes.

The eHealth platform is interfaced by a small component that receives information via a network connection and sends sensor reading into the context server. Another application interfaces a Wii Remote controller and transforms this data into a horizontal movement component.

The platform provides two components that deal with JSON as a communication mechanism. For the game, a component receives information from the context server and forms a JSON message that is sent to control the paddle or update the heart rate.

The Wii Remote interfacing software sends JSON messages about the state of the controller which are in turn read by a component in the platform. This component translates these Wii Remote events into messages for the context system.

In the following we explain in some detail for those interested how to integrate the different components by using the TZI Context Framework (see [19] for further details on the use of the framework).

6.2   JSON Adapters for the Context System

To process sensor information from JSON messages, a special component in the framework exists. The JSON Context Receiver class listens for special JSON messages on a network connection and translates them into the protocol of the context system.

It is instantiated via a Spring configuration entry as follows (using default values):

```
<bean id="jsonrec"
class="eu.rehabathome.contextdemo.JSONContextReceiver"/>
```

As an example, information from the Wii Remote interface arrives as such message: { 'ctx': { 'input': { 'wiimote': { 'hPos': {v:'0.5'} } } } }

To send JSON messages on context events, another special component exists, that waits for updates from the context server and forms appropriate messages.

The configuration defines the information to listen for and how to transform it for the receiver. In this case, an update from the Wii Remote is translated into the format understood by the game.

For every receiver needed, a new instance is created in the platform with a Spring configuration entry:

```
<bean id="jsoninput" class="eu.rehabathome.contextdemo.JSONContextSender">
        <property name="elements">
            <list>
                <bean
class="eu.rehabathome.contextdemo.JSONContextSender.Element">
                    <property name="ctx" value="input"/>
                    <property name="src" value="wiimote"/>
                    <property name="prp" value="hPos"/>
                    <property                        name="format"
value="'hPos':__VALUE__"/>
                    <property                        name="jsonFormat"
value="{__ENTRY__}"/>
                </bean>
            </list>
        </property>
    </bean>
```

6.3   Interfacing the eHealth Platform

To communicate with the e-Health platform through a network port, similar pieces of code are required to send sensor data to that network port and receive data from there.

6.4   Interfacing the Wii Remote

The Wii Remote controller from the Wii game console is a Bluetooth device. It can be added to a standard PC system by using its HID properties. HID stands for Human Interface Device and is a technical definition for interpreting input from various standard input devices such as mice, keyboards and joysticks. While the Wii Remote technically conforms to the HID specification, it is only used to allow an easier use of the device on a computer platform. Just adding the device to a system does not expose any functionality of the controller to the PC.

The HID protocol, however, allows the transmission of arbitrary data packets between the device and other software and this mechanism is exploited to interface the device. The HID protocol provides the advantage that a Wii Remote does not require a special driver on the PC side and is therefore a frequent workaround for interfacing special hardware. To access the HID layer on a Windows machine, a special software module was developed to look for Wii Remote devices and read data from them. It was tailored to the needs of the game prototype and only extracts horizontal movement information (see figure 10 for the interface).

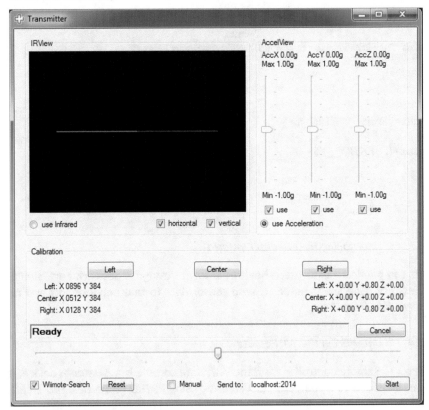

Figure 10 Wii Remote interface

While the HID layer is present on all modern operating systems, a special requirement is present on Microsoft Windows machines: The Wii Remote has to be paired to the Bluetooth stack but special software is needed for this operation.

7 Second prototype

The second prototype makes use of the Microsoft Kinect as an interaction device. Since the focus of the game is the upper body, hand tracking was chosen to control the serious game. Tracking of the hands is indirectly provided by the standard software framework that comes with the device as it provides tracking of a complete but simplified skeleton abstraction of the person in front of it. The game simply uses this information to implement a direct control scheme where visual feedback is provided, reflecting the recognized position of the hands. Using the Kinect, whilst very easy to setup for the demonstration, has a potential issue in the further development of the game: It is very suitable for tracking the movement of the hand moving horizontally and vertically but has limitations recognizing movement in the depth (e.g. moving towards and away from the monitor). Depending on the type of movement that is desired, this limitation may become a problem later.

7.1 Wii Remote vs Kinect

The Wii Remote controller needs to be held by the user of the system and also needs additional setup within the environment (e.g. the sensor-bar) it provides an easy pointing control scheme and can detect movement in depth and also rotation of the wrist. It does not however provide an abstraction of the actual pose of the user and may therefore have a more limited clinical use than the Kinect sensor. Whilst the Kinect is directly connected to the computer system via a USB connection, the Wii Remote uses a wireless connection via Bluetooth. This technology is very robust but introduces the need for an additional discovery mechanism to connect it to the software.

7.2 Abstract Game Control

One could control games directly by the software provided for the Kinect sensors. However, to be able to change the sensor systems used for input, we envision an abstraction of game controls. A potential abstraction concept replaces the direct interface inside the game with an abstract control that is interfaced by other components. See figure 11 for a conceptual sketch.

There are many benefits of control abstraction but also some downsides. If an abstraction is present, the serious game can make use of many different devices that may be added in the future. It also enforces the definition of common semantics for actions in the game, as input from other devices needs to be mapped to these meanings in order to provide appropriate control. On the other hand, making use of a device directly can provide additional information that creates a more immersive game. The body tracking provided by the Kinect for example can be used to animate a virtual representation of the user. While this is not necessary for the game, it may be a factor in acceptance and user satisfaction.

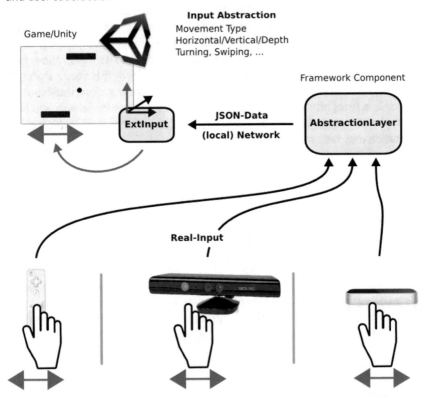

Figure 11 Control Abstraction Sketch for Device Replacement

As a proof of concept implementation, showing the potential of abstracting game input for the game engine used, a simple game-like application was developed in Unity. Instead of interfacing a game controller by using standard components, a small plugin was created that exposes a control interface to the engine. It is in turn supplied with information from outside the game engine system via a network link. An external component interfaces with an input device and transmits the data to the component inside the game engine.

In the implementation, a Wii Remote controller was used as a pointing device and the aim was used to control a game object. In the following we explain how to use a Context Framework to connect an input device and game controller. For this purpose, we use again the Wii Remote to compare a different implementation to the first prototype when using the Context Framework.

7.3   Context Framework

The Context Framework is a system to multiplex sensor information to interested clients. It is designed for low-latency transmissions with small bandwidth requirements. While it uses a specialized protocol, JSON adapters exist that make common functionality available to other components. In brief, information is divided into a *context*, a *source* and a *property*. Every property belongs to a source and every source is part of a context. There can be more than one context. For demo purposes we use again the Pong game. A context named 'input' was created where a source 'wiimote' exists. The 'wiimote' source had one property called 'hPos' that was the derived position for the Pong paddle. To enter this information into the Context Framework, an application would send a JSON message like the following to port 2014, the JSON-CTX-Bridge:

{"ctx":{"input":{"wiimote":{"hpos":{"v":"0.5"}}}}

In this case, 'v' defines the new value for the property as 0.5. There are also more property options like tagging and timestamping optional that are not covered here. When context is entered like this, it is also distributed to potential listeners: Like the JSON-CTX-Bridge there is also a CTX-JSON-Bridge that listens on the special protocol and transmits values to other components as JSON packets. This adapter can also reformat the message to suit a component better. In the Pong demo, the Wii Remote input is translated to a simpler message where just {"hPos":"0.5"} is send to the game port 12345. The translation rules and target components are configured via the context platform (Spring-Framework based configuration). Similar conversions are done for the heart-rate sensor in the demo. Java components can also make use of specialized client classes that directly interact with the Context Framework for sending and receiving information. These clients connect to the server and use the special protocol.

Messages used in the Pong Demo

The Pong Demo game name is 'pong" and this name is used in the defined messages for start-up, shutdown and keep-alive, e.g.

{"announce":"startup","params":{"game-name":"pong"}}.

Sensor input

The game uses defined message structures: For the Pong game, an input announcement can carry parameters for the heartrate - bpm - or the paddle position - hPos - or both. A full message looks e.g. like this:

{"announce":"input", "params":{"bpm":120,"hPos":0.7}},

which would set both parameters at once. But one can also send parameters separately.

Game events

The game produces two in-game events; one for the ball hitting the paddle and one for hitting the bottom.

{"announce":"game-event","params":{**"pong-event"**:"paddle-hit"}}

{"announce":"game-event","params":{**"pong-event"**:"bottom-hit"}}

7.4   Sensor Adaptor Component

The Sensor Adapter Component (SAC) sends sensor data directly to the Sensor Interpreter Component (SIC) - no routing or connection to the Control Component (CC) or other components are needed.

There are three different implementations of the SAC:

1. The Sensor Adaptor Component simulator simulates the output of a real SAC and has a Graphical User Interface (GUI) (see figure 12) that allows the user to manually control the heart rate value for testing purposes. It starts sending data after start-up without special initialization.

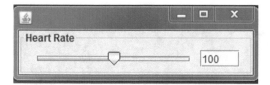

Figure 12 Heart rate simulator

2. The eHealth Sensor Adaptor Component reads sensor data from the eHealth platform (see figure 13). It simply listens on a specific port for the data from the eHealth device, extracts the heartrate and sends it to the SIC twice a second. It starts sending after start-up without requiring any special initialization .

Figure 13 eHealth platform

3. The Bluetooth Sensor Adaptor Component reads sensor data from a pulse sensor (see figure 14) connected via Bluetooth. The Bluetooth serial port is used to read the data and can be configured in a configuration file. It reads the byte stream from the Bluetooth sensor and extracts the heartrate and inter beat values. It starts sending data twice a second after start-up without special initialization. If the Bluetooth signal strength is weak, the interval between messages may increase or the sent data may be incorrect.

One can select the SAC implementation by changing the executable path in the control component configuration as described in the next section.

Figure 14 Bluetooth pulse sensor

7.5   *Control Component*

The primary class has references to all components used. Additionally, a component map exists that maps the name of the components to the corresponding object. The logic of the control component is implemented in two loops.
The first one restarts the menu component if the game is exited or crashes
while running. The second one is a busy-waiting loop of waiting for messages
and sending a keep-alive in a message loop. If a message is received by the
message loop, the message is handled in a way that the system first checks
whether a route exists for the received message in the configuration file. Then

it sends the message to the new destination. Otherwise the system uses the coded logic for messages to direct them to the Context Controller.

8 Conclusion

In this book chapter, a hardware and software approach for the Reference Rehabilitation Platform is given. This paves the way for Serious Games like those described in the next book chapter to interconnect with different input devices and sensors provided by the eHealth platform. The advantage is the flexibility with respect to the chosen devices as games. Standardization is given without knowing in advance any requirements with respect to the desired therapy, appropriate medical data and therapeutic activities being monitored.

9 References

[1] Difede, J. & Hoffman, H. (2002). Virtual reality exposure therapy for World Trade Center Post Traumatic Stress Disorder. Cyberpsychology and Behavior, 5:6, pp. 529-535

[2] Ready, D. J., Gerardi, R. J., Backscheider, A. G., Mascaro, N., & Rothbaum, B. (2010). Comparing Virtual Reality Exposure Therapy to Present-Centered Therapy with 11 U.S. Vietnam Veterans with PTSD. Cyberpsychology, Behavior & Social Networking, 13(1), pp. 49-54.

[3] V. Popescu, G. Burdea, M. Bouzit and V. Hentz, "Ortho-pedic telerehabilitation with virtual force feedback,"IEEE Trans. In-form. Technol. Biomed., vol. 4, pp. 45–51, Mar. 2000.

[4] S. K. Numrich, "Culture, models, and games: Incorporating warfare's human dimension," IEEE Intell. Syst, vol. 23, no. 4, pp. 58–61, 2008.

[5] C. G. Von Wangenheim and F. Shul, "To game or not to game?," IEEE Softw., vol. 26, no. 2, pp. 92–94, 2009.

[6] M. Macedonia, "Virtual worlds: A new reality for treating post-traumatic stress disorder," IEEE Comp. Graph. App, vol. 29, no. 1, pp. 86–88, 2009.

[7] B. Sawyer, "From cells to cell processors: The integration of health and video games," IEEE Comp. Graph. App, vol. 28, no. 6, pp. 83–85, 2008.

[8] P. Rego, P. Moreira, and L. Reis, "Serious games for rehabilitation: A survey and a classification towards a taxonomy," Information Systems and Technologies (CISTI), 2010, pp. 1-6.

[9]   J. Wiemeyer and A. Kliem, "Serious games in prevention and rehabilitation—a new panacea for elderly people?", European Review of Aging and Physical Activity, vol. 9, no. 1, pp. 41–50, Dec. 2011.

[10]   M. Pirovano, R. Mainetti, G. Baud-bovy, P. L. Lanzi, and N. A. Borghese, "Self-Adaptive Games for Rehabilitation at Home," in IEEE Computational Intelligence and Games (CIG), 2012, pp. 179–186.

[11]   M. Ma and K. Bechkoum, "Serious Games for Movement Therapy after Stroke," IEEE Int. Conf. on Systems, Man and Cybernetics, pp. 1872–1877, 2008.

[12]   J. W. Burke, M. D. J. McNeill, D. K. Charles, P. J. Morrow, J. H. Crosbie, and S. M. McDonough, "Optimising engagement for stroke rehabilitation using Serious Games," The Visual Computer, vol. 25, no. 12, pp. 1085–1099, Aug. 2009.

[13]   A. Conconi, T. Ganchev, O. Kocsis, G. Papadopoulos, F. F.- Aranda, and S. Jiménez-Murcia, "PlayMancer: A Serious Gaming 3D Environment," in Conf. on Automated Solutions for Cross Media Content and Multi-channel Distribution - AXMEDIS'08, 2008.

[14]   N. A. Borghese, R. Mainetti, and P. L. Lanzi, "An Integrated Low-Cost System for At-Home Rehabilitation," in Virtual Systems and Multimedia (VSMM), 2012, pp. 553–556.

[15]   Hasomed GmbH, "RehaCom basic manual," 2009.

[16]   M. S. Cameirão, S. B. Badia, L. Zimmerli, E. D. Oller, and P. F. M. J. Verschure, "The Rehabilitation Gaming System: a Review," Studies in Health Technology and Informatics, vol. 145, pp. 65–83, 2009.

[17]   R. Mainetti, A. Sedda, M. Ronchetti, G. Bottini, and N. A. Borghese, "Duckneglect: Videogames based neglect rehabilitation," Technology and Health Care, vol. 21, no. 2, pp. 97–111, 2013.

[18]   Alankus, G., Lazar, A., May, M., Kelleher, C., 2010. Towards customizable games for stroke rehabilitation. Proceedings of the 28th international conference on Human factors in computing systems (Atlanta, Georgia, USA, 2010), pp. 2113-2122.

[19]   H. Iben, 'Rapid Prototyping Infrastructure for Wearable Computing Applications', Dissertation University Bremen, 2015, https://elib.suub.uni-bremen.de/edocs/00105034-1.pdf

Acknowledgment

The authors of this chapter thank all their co-workers and funding agencies of the many projects in recent years. Without this support, this chapter could not exist.

IV Serious Games for Neuro-Rehabilitation A User Centred Design Approach

Antonio Ascolese / Lucia Pannese / David Wortley

Abstract

In this book chapter, we describe a set of Serious Games suitable for therapeutic purposes in Neuro Rehabilitation. This set of games was developed using the Reference Rehabilitation Platform (RRP) for Serious Games as described in book chapter two. The design and development of these games follows a user centred design approach. This book chapter provides an insight into the development cycles, including the results from evaluations with patients and stakeholders of rehabilitation clinics. The games are designed for the rehabilitation of the upper extremities although the RRP allows extensions towards the whole body, fine hand movements, as well as cognitive and aphasia training; the modularity of the RRP concerning the sensor integration enables this. Although the games described here mainly used the Kinect as input device, the context framework approach described in book chapter three allows an extension towards other input devices using the eHealth platform described there.

1	Introduction
2	Requirements
3	Development of Technical Solutions for Rehabilitation
4	Case History: The Development of Rehab@Home
5	Conclusion
6	References

© Springer Fachmedien Wiesbaden GmbH, part of Springer Nature 2018
M. Lawo und P. Knackfuß (Hrsg.), *Clinical Rehabilitation Experience Utilizing Serious Games*, Advanced Studies Mobile Research Center Bremen, https://doi.org/10.1007/978-3-658-21957-4_4

1   Introduction

Even if the 'serious game' concept is nowadays well established, a consistent and generally accepted definition of this term has not yet been agreed upon. According to Ben Sawyer [1], a serious game is 'any meaningful use of computerized game/game industry resources whose chief mission is not entertainment': however, it is possible to define a serious game as an interactive simulation which has the look and feel of a game, but is actually a simulation of real-world events or processes or is represented in form of a metaphor. Over the past few years, with the widespread use of commercial games, the domain of game-based learning has received increasing attention.

With these advances in technologies and revolutions in patients', trainees', and public expectations, the global healthcare sector is increasingly turning to Serious Games to solve problems [2]. Serious Games are in fact an emerging technology growing in importance for specialized training, taking advantage of 3D games and game engines in order to improve the realistic experience of users. They are so popular because of the availability and ubiquity of devices, easy-to-use interfaces, attractive highly motivational graphics and sounds, and an ability of self-paced, entertaining learning through role-playing and simulation. Therefore, the optimal adoption of Serious Games should consider learners' specific ways of learning in order to achieve an increase in nutritional knowledge and a promotion of physical activity as means of preventing obesity.

Furthermore, over the past few years, there has been an increased use of digital technologies, including games, to initiate and sustain engagement across the healthcare sector, where the focus on tackling the attitudes and behaviours that may lead to complicated health conditions has seen games deployed to wide audiences ([2], [3] and [4]). Thus, according to the advancement of technology and the desire to achieve good health in an interesting and enjoyable way, different Serious Games for health particularly targeted at kids and teens have been proposed during the last few years [5].

According to several studies (i.e. [4], [6] and [7]), the issue of individuals' motivation to change is the most significant obstacle to promoting positive health behaviour. The ability of games-based activities to reach and engage large number of players for long periods of time provides an opportunity for them to be used as a pedagogical tool. Activities related to healthy living require individuals to embrace delayed satisfaction, where the reward may be as obscure as the prevention of a chronic condition [3]. The use of game mechanics and concepts to facilitate participation in such activities will commonly benefit from rewards and incentives used to sustain positive engagement. Behavioural change may be initiated by extrinsic sources of motivation or external factors that influence how we behave [8]. The long-term goal will include promoting intrinsic motivation and positive habits through sustained engagement, where individuals could discover their personal incentives and rewards for healthy behaviour. Both the use of digital games and the concept of gamification present, in fact, an opportunity for better understanding of individuals' knowledge, attitude and behaviour and assessment of their progress, the provision of more personalised feedback and support towards a healthier lifestyle.

2 Requirements

Patients, therapists and the health system have different requirements concerning Serious Games. Where motivation, quality of life and sticking to a therapy are important for the patient, it is the availability of multiple datasets, the customization of the rehabilitation program to the individual patent's needs and the balance of clinical necessity and fun for the therapist. For the health system, the reduction of costs is predominant; however aspects of an empowering system and the complementarity to rehabilitation in general are further issues.

2.1   Patient

Games are made to involve players, so the main targets of Serious Games in the rehabilitation sector are patients. Various diseases require rehabilitation to recover from specific damage: in this context, exercises can be repetitive and each patient has to improve by completing them repeatedly and independent of external help. Hence, games show promise for helping patients gain a central role in the rehabilitation process.

Motivation

The motivation mechanics underlie the concept of gamification itself ([9] and [10]). If participants are not motivated to engage with a system/app, they may not experience the full benefit of the proposed solution. It is also important to address participants' motivations to engage with the technology used in a specific context, as unfamiliarity and feelings of incompetence could be a barrier to participation. The most important tips to follow in order to increase the users' engagement in gamified apps, is to 'intrinsically motivate' users to participate in the gamified solution. In general, motivation refers to psychological processes that are responsible for initiating and continuing goal-directed behaviours [11]. In particular, intrinsic motivation is defined as doing an activity for its inherent satisfaction rather than for some reward which is not directly linked to the pleasure of completing the activity. When intrinsically motivated, a person is moved to act for the fun or challenge entailed rather than because of external requests, pressures, or rewards. This happens because intrinsic motivation is based on the human need to be competent and make choices without external influence (according to the Self-Determination theory of Deci and Ryan, [12]). An example of intrinsically motivated behaviour is the amount of time that people dedicate to hobbies such as sport, playing computer games or going to the theatre despite the fact that these kinds of activities do not bring any kind of tangible reward to the participants.

This type of motivation differs from the extrinsic one, which instead refers to doing something because it leads to a separable outcome (i.e. points, badges or feedback). An example of extrinsically motivated behaviour is the time and effort that a student dedicates to study in order to get a good grade at school.

Despite the differences underlined before, the intrinsic and extrinsic motivators are not rigidly separate entities. Deci & Ryan [12] argue that these motivators are fluent, and that by offering extrinsic rewards (i.e. a prize) for the completion of a specific behaviour and/or for the reaching of a goal, which is meaningful and pleasurable, it is possible to satisfy the intrinsic human need for competence and self- determination. In this way, people can start completing an activity that was not initially interesting or inspiring because it is fun, not simply because they are being rewarded. This happens because users adopt the extrinsically motivated behaviour as though it were intrinsic.

With respect to the 'health-related behaviours', in general, users are not intrinsically motivated to adopt them. Since humans like to feel in control, simply offering extrinsic rewards when gamifying health behaviours can diminish users' internal drive to complete an activity ([13] and [14]).

Quality of Life

The gaming solution developed for neuro-rehabilitation should be portable, ideally using technological devices which can be easily and affordably purchased or even already available in the daily life of patients. Several advantages are involved in this approach to rehabilitation, mainly affecting the quality of life since patients can avoid a substantial change in personal habits while following the rehabilitation program.

Firstly, using home-based technological solutions avoids the need to travel to the clinic, which, in the case of rehabilitation is not likely to be a single trip, but will have a specific routine - for example twice a week. Due to the fact that most diseases deal with motor impairments, traveling to the clinic can be very strenuous for the patient and it probably requires the involvement of different caregivers, also bringing a perceived lack of autonomy. Furthermore, some patients may not have the choice because their level of impairment or lack of a strong social network makes them unable to get to a hospital.

Secondly, the presence of a familiar home environment during the rehabilitation process represents an added value: the patient feels more comfortable and thus the initiative is taken more frequently and personal goals are more likely to be achieved than in hospital context [15]. This factor is emphasized for the elderly because they are more comfortable in a home environment and experiencing improvement in a real context may enhance personal self-efficacy about rehabilitation tasks [16].

Compliance: Sticking to Therapy

Motivation and the presence of a familiar environment have, as a common aim, strengthening the involvement of the patient in the rehabilitation program which leads to a stronger rehabilitation outcome; clearly, efforts to develop portable gaming solutions will only be of value if the final medical results of the patient show some improvement.

The main variable in the process is therapeutic compliance, because exercising for the right amount of time and in the correct way is intuitively the strongest predictor for a better recovery. Specific game mechanics can positively affect compliance, for example specific feedback about performance and the facility to check progress. Feedback can be shown to users in real-time by tracking patients' physical movements with lines on the screen so that the patient can see a trace of the real effort. Auditory and/or visual cues can be used to emphasise a positive outcome of a task. Scores are important to immediately show how well the patient is training and they are also the main mechanism through which it is possible to track personal progress. Showing a record of daily scores is a simple yet powerful way to show to the patient that their performance is improving session by session. Both these mechanics can illustrate the value of systematic home-based training as a structured path where the patients can visualise the steps needed to reach the final rehabilitation goal.

2.2   Therapist

Despite the fact that the main aim is to train patients and improve the efficacy of rehabilitation, therapists also have specific benefits in working with Serious Games: they do not play nor develop the solution, since their crucial role is in working on game settings and customization of rehabilitation programs. When designing a gaming solution, the role of the therapist is necessary because there is the need to create the right set of opportunities necessary for the various clinical cases.

Multiple Dataset Availability

Home rehabilitation may appear to present barriers to providing patient progress feedback to clinicians if training takes place in a setting remote from the therapist's workplace. However, the technological devices needed to practice in home environment automatically collect multiple sets of objective data that are, paradoxically, not taken into account when rehabilitation is set in a hospital dealing with classical exercises. In nursing contexts, therapists use professional experience to analyse the performance of patients and this is undoubtedly a powerful source of knowledge; on the other hand, numeric data offered by technology can provide a wider and more objective source of knowledge to check the performance and the clinical status of the patient. Therapists can be trained to acquire the knowledge they need by referring to the database and at the same time, other medical data not typically relevant to the task, such as heart rate, can be made available and thus monitored to provide additional feedback about the health status of the patient.

Customization of the Rehabilitation Program (Adaptability)

The amount of data available provides a deeper knowledge of each patient so that it becomes possible to adapt the rehabilitation training to the specific needs of every single case. Serious Games already provide a set of levels and the ability to calibrate the complexity of the tasks scheduled so that the difficulty levels can be automatically set on the basis of players' abilities. Each patient has a personal program which automatically updates when player's performance improves and exercises potentially start to be boring. The customization of the rehabilitation program can be either automatically made through technology or manually through the personalized options that therapists have access to, based on personal and detailed feedback available about performance and clinical condition.

Compliance: Balance of Clinical Necessities and Fun

As outlined in the previous paragraph, compliance is a key issue for patients and it is undoubtedly the same for therapists, who usually have to balance the clinical necessities of multiple repetitions and boring movements with other activities that can be more enjoyable for the patient. Serious Games help to provide this balance thus avoiding premature dropouts that can erode the initial progress made. The most important tasks can be scheduled in spite of their boredom because the game environment and dynamics help to make them more enjoyable. Moreover, the reports on game performance are an important source for therapists to verify the actual execution of the training program and track the progress made.

2.3   Health System

When technological solutions and Serious Games are implemented by health institutions, even if the final cost burden – or part of it – is borne by the patients, the link to a clinical centre is necessary. Therefore, it is clear that serious games should provide added value not only for patients and therapists but also for the health system they are connected to. In this way, home based rehabilitation using Serious Games can create an ecosystem of win-win relationships between patients, therapists and health systems.

Reduction of Costs

The big challenge for health systems is the reduction of costs. Serious Games can address this task, mainly by empowering users with the ability to complete part of their rehabilitation alone, without the continuous assistance of medical manpower. This lack of professional assistance does not mean a decreased need for therapists, but creates opportunities for therapists to focus on those situations that require deeper medical attention, e.g. for patients who cannot leave the hospital because of poor overall clinical condition. In this way it is possible to optimize the efforts of key personnel, without tying up intensive care resources for patients that could practice rehabilitation autonomously. It is also likely that therapists could also be more motivated, removing the need to do repetitive low intensity programs and enabling remote supervision via technological solutions.

Empowering Systems

Technological solutions like Serious Games have an intrinsic capacity to adapt to different clinical conditions and different patients, as already explained. Simultaneously, it is possible to collect big amounts of data and identify common trends around populations, i.e. a disease, or specific contexts. In this way, each care centre could improve the knowledge of their users through this statistical data and thereby enable them to improve the quality of the clinical offer. Serious Games can evolve into a learning system which stimulates the development of further mini-games and personalisation to better fit the target audience.

Many Serious Games facilitate the creation of web-based communities around them, consisting of the different players involved: these communities are supported by the sharing of game benefits, the ability to compare results and possible collaborative game dynamics implemented by specific games. Internal messaging systems are often provided for connecting to other players and this, in the context of medical applications, can prompt the creation of specific communities of patients sharing the same clinical condition with the possible formation of groups based on self-help dynamics.

Compliance: From Surgery to Final Recovery (Complementary to Rehabilitation - 2nd Phase)

In the rehabilitation phase, the aim of clinical centres is to achieve the goal of a complete recovery from damage, as far as is possible. For example, it is likely that, after a stroke, some functionality can no longer be reactivated. Despite this, it is still possible to identify a level of functioning that allows the patient to live as full a life as possible. If this can be realized, the patient is likely to require fewer accesses to hospital resulting in consequential benefits, such as the reduction of the waitlist and the ability to focus on clinical cases needing high frequency care. Hence, the efficacy of Serious Games in enhancing clinical compliance can have benefits for health systems both because of the satisfaction perceived by patients who have a better outcome and because of the lowering of the clinical burden on staff and resources.

In summary, the central idea is that Serious Games can enhance the compliance of patients which is a crucial objective for everyone involved in the rehabilitation process and can thereby offer specific benefits at different levels. Firstly, games can ensure that patients stick to their therapy program. Secondly, therapists can leverage this compliance to suggest additional tasks by exploiting the data collected by the technology in the home environment. Finally, achieving the goal of a more complete rehabilitation will surely have a positive impact on the overall health system.

3 Development of Technical Solutions for Rehabilitation

In order to design and develop Serious Games that are meaningful, pleasurable and relevant for the target users, especially with respect to the health sector, a user-centered design (UCD) approach is highly recommended, aimed at applying game mechanics appropriately for the specific target population.

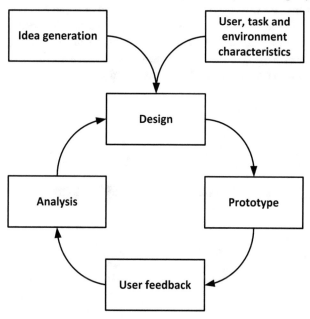

Figure 1 User centered design approach [17]

3.1 Involving User (UCD)

UCD relies on users' involvement during all the phases of the design process. Starting from the beginning of the design activities, end users are involved in several focus groups or interviews in order to understand what their needs and expectations are. Their involvement is continuous during the design of the tool until it is considered acceptable by end users (see figure 1).

This approach is one of the most suitable for the design of tools [18], [19] because

- it allows for the development of a tool that is both usable and accepted by users;
- it aims to reduce the cost and effort of fixing usability problems;
- it creates a product offering a more efficient, effective, satisfying, and user-friendly experience.

It can also help to avoid the risk of designing something that is in line with the project objectives but not interesting or effective for the end users. Users' feedback has to be integrated with therapists' guidelines to develop proto-types. The special needs of the target group of elderly people with only limited technical experience require further attention. The book chapters six and seven address these challenges from a non-technical perspective.

3.2   Basic Prototype Development

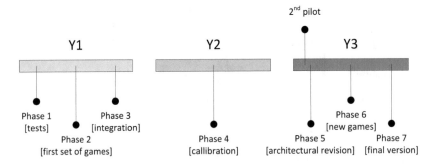

Figure 2 Basic prototype timeline

A specific sequence of design must be followed to create prototypes of game-based platforms, scheduling both implementation and field testing phases: Figure 2 above shows as an example the timeline for the development of the Rehab@Home games over the project period of three years. Book chapter three covered the hardware related aspects, here we emphasise the phases of development with users.

The development has seven phases with specific targets on the way to the final version:

- *Phase 1*: the purpose of the first version of the prototype is to perform technical tests with the device in order to understand how it works, for example to understand the real functioning of e.g. the Unity Kinect Plugin (Kinect with MS-SDK), and to test some interaction between input trackers, virtual avatar, and virtual objects (see figure 3).

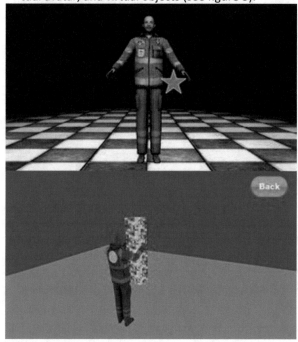

Figure 3 Preliminary tests in Phase 1

- *Phase 2*: the second version of the basic prototype includes the first version of the Serious Games developed; in this prototype all the interaction with e.g. the Microsoft Kinect device are handled using a virtual avatar.
- *Phase 3*: in the third version of the basic prototype the Serious Games are integrated within a single scene handled by a main controller script, and the use of the virtual avatar is limited to the first game; the interaction between Mcrosoft Kinect device and the interface, and the interaction between the device and the interactive objects of the second and third game are now handled directly by a cursor tracker.
- *Phase 4*: the fourth version of the basic prototype implements major interface changes and the calibration scene, following the indications of trials

with patients, such as the size and positioning of buttons. Feedback on animations and sounds are also integrated.

- *Phase 5*: the fifth version of the prototype is a crucial step in the development of the enhanced prototype, implementing some minor changes in the interface and fixing some bugs into a clearer architectural view. Each game becomes an autonomous process that can be easily added in a global recognizable structure.
- *Phase 6*: new version of games can be added if patients need supplemental exercises.
- *Phase 7*: final changes requested from the therapists based on patients' behaviour are carried out. Technical compatibility with upgraded versions of the system is checked as well as the integration with the other components of the patient station and the backend system (doctor station).

4   Case History: The Development of Rehab@Home

In the Rehab@Home project the aim was to systematically explore the design and implementation of gaming platforms for rehabilitation. These gaming platforms should improve the quality of rehabilitation process by adding further rehabilitation possibilities to existing stationary and ambulant rehabilitation. Specifically, personalized, non-intrusive, continuous and connected rehabilitation experience for patients through interactive gaming with pervasive sensing technologies were targeted.

The aim was also to reduce the manpower required for monitoring and coaching individual patients at rehabilitation centres. Through the integrated rehabilitation gaming system, heterogeneous physical and biological data should be collected, processed and communicated for clinic analysis, self-monitoring programs and system improvements.

The Rehab@Home gaming environment was meant to provide support during the rehabilitation process of users with physical or cognitive injuries, in particular post-stroke patients or those suffering from multiple sclerosis. The basic idea was to encourage the users to perform specific movements, suitable for their rehabilitation process, by playing Serious Games on consumer technologies.

The Rehab@Home gaming environment is made of different sets of Serious Games, each with different movements to perform and different purposes. The first set used the Microsoft Kinect device to exercise arms, shoulders and hands of the users through three Serious Games with the following therapeutic exercises:

- The *first game* focuses on the coordinated movements of both arms. The user controls a virtual avatar and needs to interact with virtual objects that are falling from the top of the screen.
- The *second game* focuses on using and stretching a single arm. The user must move a virtual object from the centre of the screen to four different positions, according to its colour. This game also exists in another version in which the user also has to grab the object.
- The *third game* focuses on controlling and performing small movements with a single arm. Here, the user must move a virtual object along a narrow path.

The game trackers on the screen follow the hands of the user as detected by the Microsoft Kinect in order to let all the users to play the game without problems regardless of potentially limited body mobility. In order to achieve this, the Rehab@Home game platform includes also a calibration section.

4.1 Prototype Evaluation

For the development of Rehab@Home, patients were selected according to the following inclusion criteria:

- Pathology: post-stroke, multiple sclerosis or Parkinson's disease patients;
- Minimum Active ROM[1] shoulder: 45°;
- Minimum Active ROM elbow: 45°;
- Ability to understand requests (according to the health professional);
- Lack of other diseases (i.e. orthopaedic, psychological, etc.) that could interfere with the exercise execution.

The trial was carried out using a living lab approach. - A living lab represents a user-centric research methodology for sensing, prototyping, validating and refining complex solutions and evolving real life contexts [20]. - In our trial, twelve rehabilitation sessions (covering a month period) were organized for each patient, in order to have the novelty effect wear off and detect more interesting aspects related to user acceptance and the adoption process. - This

---

[1] ROM: range of motion

more long-term user testing allows to understand whether the gaming environment and input-output set ups are good enough to motivate patients to comply with the rehabilitation therapy in a sustainable way over time. It also provides an understanding of the kind of problems which may arise when deploying the Rehab@Home solution in real-life, everyday home environments.

Twenty patients (ten from each of the two clinics in Austria and Italy) with motor impairments of the upper body were involved in the trial within the mentioned twelve individual sessions of up to one hour of duration in addition to the 'traditional' rehabilitation program. Twenty patients (control group) with the same inclusion criteria followed the same steps, except for the use of the Rehab@Home solution. The patient received an Information Sheet and Consent Form to sign and had to complete a Pre-Trial Questionnaire, including some Patient Profiling questions, the Motivational Index (short version), ICF[2] assessment, and Intention to Use question. The patients performed the twelve sessions with the Rehab@Home patient station, filing a Post-Session Questionnaire at the end of each, with questions about user experience and satisfaction. The Post-Trial Questionnaire, again included the Motivational Index (short version), ICF assessment, and Intention to Use question.

The difference between this and the Pre-Trial Questionnaire is the shift in focus from a user profile to a user intention to use.

The preliminary results of Rehab@Home involved 15 persons (mean age 58.73, SD 12.78; post-stroke: N = 5, Multiple Sclerosis: N=10) in the pilot study of therapeutic effectiveness, completing the whole treatment of twelve sessions.

Results showed that there was an improvement in functional abilities and fine hand use (Box and Block Test, Nine hole peg test)[3]. Further evaluation of health perception (EQ 5D-5L)[4] and the participants perception of wellbeing (Short Form 12)[5] revealed an improvement in those domains following rehabilitation with Serious Games. These results were corroborated by reduction in the severity of impairment and activity limitations as classified through the ICF core set. In general, differences observed through the ICF core set following rehabilitation were small, but all showed improvements in the various do-

---

[2] ICF: International Classification of Functioning, Disability and Health
[3] See book chapter eight for details
[4] https://euroqol.org/eq-5d-instruments/eq-5d-5l-about/
[5] https://www.rand.org/health/surveys_tools/mos/12-item-short-form.html

mains. As an example, Exercise Tolerance increased enough to change the qualifier status from very limited exercise tolerance to moderate impairment of exercise tolerance. Moreover, fine use of hand was also improved.

Overall, it can be concluded that the devices and games proposed to participants were positively accepted. A patient feedback, a week after he finished the protocol: *'I would do another 40 sessions if I could. I believe it helped me a lot!'*

4.2 Description of the Final Solution

The Rehab@Home game was finally developed using the Unity3D game engine and the Microsoft Kinect and will be described in the following section. The game was divided into two different scenes: Calibration Scene and Game Scene. A message function is also available, located on the homepage of the Rehab@Home patient station as an initial alternative to the games section when starting to use the service.

Calibration Scene

The Calibration Scene handles the calibration to allow users with limited body mobility to play the games. The users must try to reach four signs placed in the corners of the screen. The script records the nearest position reached for every corner and then calculates the maximum area that the user can reach based on those coordinates. This 'calibration rectangle' is used to adapt the information about the tracker's position. The information about the calibration rectangle is passed to the Game Scene.

The calibration process provides a way to determine the maximum area in the space reachable by the actual user with limited body mobility and according to limit the workspace based on those limitations. Using Microsoft Kinect, the 3D space defined by the user's movement is converted into a screen position according to a viewport of size 1 by 1, which is used in all games' mechanics. The calibration modifies this viewport area using four parameters: x and y define the translation from the centre of the viewport, while width and height define the size of the new viewport area.

Figure 4 Calibration – 1st and 2nd phases

This process consists of two steps (see figure 4): in the first one, where the users have to try to reach four signs placed in the corners of the screen, the algorithm records the closest position reached for every corner and then calculates the maximum area that the user can reach based on those coordinates, modifying the four parameters (x, y, width and height) seen before.

The second step is a test for this goal in which the user must actually reach four signs placed in the corners of the screen to complete the calibration process. If for some reason a user is unable to reach one of the signs, it is possible to move back to step one to perform a new calibration using the Reset button.

Games Description

The Game Scene is divided into four parts, the main controller area and a part for every game (see figure 5). The main controller handles the interactions between the game and the Microsoft Kinect device, controls the main menu of the application, and starts every single game. Once the user chooses the game, the main controller identifies the right object called 'game logic' and activates it to relinquish the control to the game controller. Every game controller is active during the game lifecycle (setup, start, game, pause, and end) and contains all the procedures and information about the game operation. Besides the 'Rest Game', in total six games were developed as described below: 'Bees and Flowers', 'Popping Flowers', 'Coloured Cans', 'Grab Your Can', 'Blackboard' and 'Mad Fridge'. All games are without an avatar.

Figure 5 Games menu

- Rest Game

Actually, 'Rest game' is not a game (see figure 6). It is just a three minutes video included at the beginning of each Rehab@Home session to relax users and record a baseline level of the physiological parameters.

Figure 6 Rest Game

- Bees and Flowers

The objective of this game is to touch correct items (flowers) and avoid wrong ones (bees). The context is a garden. Items fall down from the top (see figure 7). The instructions for the user at the beginning and during the game are: 'Touch as many flowers as you can and avoid the bees'. The Microsoft Kinect is used to monitor the movement of one hand. Visual and two different auditory feedbacks are given based on the touched item. As scores +100 is given for correct items and -50 for wrong items as calculated during the session.

Figure 7 Bees and Flowers game

Input data is selected using the doctor station: The *duration* indicates how long a game lasts, the *velocity* how fast objects fall through the screen or how much time objects stay on screen, and the *density* defines the maximum number of objects that can be active on the screen at any instant. Together with velocity, density states the minimum time between each spawning of a new object. The *proportion* states the percentage of wrong objects from all objects spawned.

- Popping Flowers

The objective of the 'Popping Flowers' game is to touch correct items (flowers) and avoid wrong ones (bees), where items appear all-around the screen (see figure 8). The instructions available at the beginning and during the game are: 'Touch as many flowers as you can and avoid the bees'. The Microsoft Kinect is used to monitor the movement of one hand. Visual and two different auditory feedbacks are given based on the touched item. Scoring and input data are the same as in the previous game.

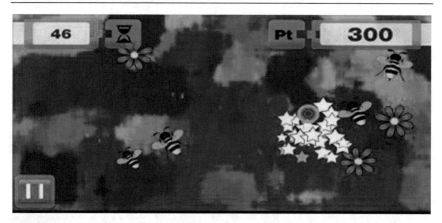

Figure 8 Popping Flowers game

- Coloured Cans

The context of this game is the kitchen. The objective is to move cans to the correct place according to colours, e.g. the red one right at the top, the blue one bottom left, green under the table, yellow on the table, etc. (see figure 9). The instructions for the user at the beginning and during the game are: 'Move the can to the correct shelf'.

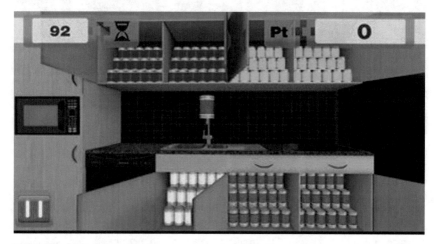

Figure 9 Coloured Cans game

The Microsoft Kinect is used to monitor the movement of one hand. Visual and two different auditory feedbacks are given based on the place where one puts a box, it explodes or not. As scores ±100 is given for correct or incorrect actions as calculated during the session.

Input data is selected using the doctor station: The *duration* indicates how long a game lasts, the *number of camps* refers to how many different colours of cans will be used in the game, and which subdivision of colours among shelves have to be used. The *precision* determines the size of area the target shelves of the game occupies.

- Grab Your Can

The game is similar to the previous one. However, the cans are moving (see figure 10). The input data are the same as those of the 'Coloured Cans' game.

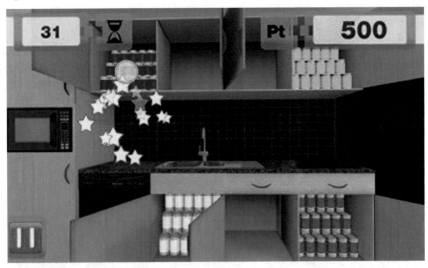

Figure 10 Grab Your Can game

- Blackboard

The context of this game is the kitchen blackboard. The shape on the left hand side one has to move to coloured spots on the right hand side by following a random path (see figure 11). Random pairings are proposed (e.g. star-blue, square-red). Red dots that appear along the path should be collected (see figure 9). The instruction for the user at the beginning and during the game is: 'Move the shape on the left hand side to the correct colour on the right hand side through all the red circles'. The Microsoft Kinect is used to monitor the movement of one hand. Visual and two different auditory feedbacks are given based on the correct association of shape and colour. As scores + 100 for the correct shape, +10 for each collected dot, -10 for each time users get wrong trajectory.

Input data (selected by the doctor station) are the *duration* defining how long the game should last and the *trajectory dimensions* that define the tolerance space around the trajectory that users have to try to follow.

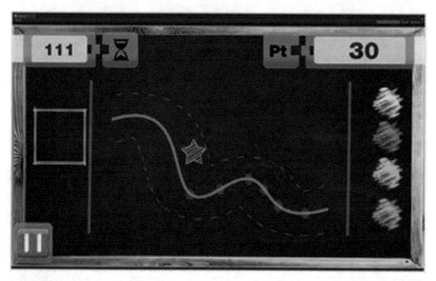

Figure 11 Blackboard game

- Mad Fridge

The context of this game is the kitchen table. One has to collect eggs and avoid gears cast away from the fridge, moving the basket on the table (see figure 12). The instruction for the user at the beginning and during the game is: 'Collect as many eggs as you can and avoid the gears'. The Leap Motion or the Microsoft Kinect are used to monitor the movement of one hand. Visual and two different feedbacks are given based on the correct object collected, avoided or missed by the basket. As scores + 100 for the correct items and -50 for wrong items the user receives during the session.

Figure 12 Mad Fridge game

Input data is selected using the doctor station: The *duration* indicates how long a game lasts, the *velocity* how fast objects fall through the screen or how much time objects stay on screen, and the *density* defines the maximum number of objects that can be active on screen at any instant and indicates together with velocity, the minimum time between each spawning of a new object. The *proportion* states the percentage of wrong objects from all objects spawned. The *portion of table* defines which target points should be active in the game and the *basket dimension* is the size of the basket.

5   Conclusion

To be effective, rehabilitation has to be early, intensive and repetitive, especially for some clinical conditions (e.g. stroke). It can be a lengthy process whose success depends on patient's endurance and discipline. Serious Games using readily available and affordable consumer technologies open up the opportunity to provide effective rehabilitation for a variety of patient conditions in a home environment.

We explored the potential of these technologies to create a new approach to patient rehabilitation designed to deliver the following benefits:

- Increased effectiveness as an approach to therapy by virtue of the inherent ability of games psychology and mechanics to provide motivation and incentives to persist with treatment;
- The importance of a home environment, especially in certain cases, to make therapy as accessible and affordable as possible;
- The reduction of therapy costs both to the patient and the health service by delivering therapy at the patient's home rather than a hospital or clinic;
- The ability of the therapist to access data on patient's progress remotely, to adjust the exercise program in a personalized way and to enable them to focus their time and attention on more complex patient cases.

The pilot project results provided encouraging signs that games-based therapy offers similar outcomes and improvements in patient mobility to traditional approaches but at potentially significantly reduced costs. The evidence of the trials also showed a positive acceptance of these games by the patients.

The Ageing Society and physical and cognitive impairment represent a global problem which cannot be addressed within existing health service structures and costs and the use of games technologies. Gamification strategies and enabling technologies to address these health challenges become increasingly important.

6   References

[1]   Sawyer, B. (2004). The Serious Games summit: emergent use of interactive games for solving problems is serious effort. Computers in Entertainment (CIE), 2(1), p.5-5.

[2]   Baranowski, T., O'Connor, T., Hughes, S., Beltran, A., Baranowski, J., Nicklas, T. & Buday, R. (2012). Smartphone video game simulation of parent-child interaction: Learning skills for effective vegetable parenting.Serious games for healthcare: Applications and implications. Hershey, PA: IGI Global, pp. 248-265.

[3]   Arnab, S., Brown, K., Clarke, S., Dunwell, I., Lim, T., Suttie, N. & De Freitas, S. (2013). The development approach of a pedagogically-driven serious game to support Relationship and Sex Education (RSE) within a classroom setting. Computers & Education, 69, pp.15-30.

[4]   Ulicsak, M. (2010, October). You can learn Your Parents are Immature: An Analysis of What Learning can Result from Family Video Gaming. In

Proceedings of the 4th European Conference on Games-Based Learning: ECGBL2010 (p. 403). Academic Conferences Limited.

[5]   Wattanasoontorn, V., Boada, I., García, R., & Sbert, M. (2013). Serious games for health. Entertainment Computing, 4(4), pp. 231-247.

[6]   Baranowski, T., Buday, R., Thompson, D. I., & Baranowski, J. (2008). Playing for real: video games and stories for health-related behavior change. American journal of preventive medicine, 34(1), pp. 74-82.

[7]   Seifert, C. M., Chapman, L. S., Hart, J. K., & Perez, P. (2012). Enhancing intrinsic motivation in health promotion and wellness. American Journal of Health Promotion, 26(3), TAHP-1.

[8]   Kato, P. M., Cole, S. W., Bradlyn, A. S., & Pollock, B. H. (2008). A video game improves behavioral outcomes in adolescents and young adults with cancer: a randomized trial. Pediatrics, 122(2), pp. e305-e317.

[9]   Deterding, S., Dixon, D., Khaled, R., & Nacke, L. (2011). From game design elements to gamefulness: Defining gamification. Proceedings of the 15th International Academic MindTrek Conference: Envisioning Future Media Environments, pp. 9-15.

[10]   Wu, M. (2011). Gamification 101: The psychology of motivation. Lithium Community, 4.

[11]   Schunk, D.H., Pintrich, P.R., Meece, J.L. (2010). Motivation in education: theory, research, and applications. Pearson, Upper Saddle River.

[12]   Deci, E. L., & Ryan, R. M. (1985). Intrinsic motivation and self-determination in human behaviour. New York: Plenum Press.

[13]   Deci, E. L., Koestner, R., & Ryan, R. M. (1999). A meta-analytic review of experiments examining the effects of extrinsic rewards on intrinsic motivation. Psychological Bulletin, 125(6), p. 627.

[14]   Nicholson, S. (2012, March). Completing the experience: Debriefing in experiential educational games. In Proceedings of the 3rd international conference on society and information technologies, pp. 117-121.

[15]   Koch, L. V., Wottrich, A. W., & Holmqvist, L. W. (1998). Rehabilitation in the home versus the hospital: the importance of context. Disability and rehabilitation, 20(10), pp. 367-372.

[16]   Sanford, J. A., Griffiths, P. C., Richardson, P., Hargraves, K., Butterfield, T., & Hoenig, H. (2006). The Effects of In-Home Rehabilitation on Task Self-Efficacy in Mobility-Impaired Adults: A Randomized Clinical Trial. Journal of the American Geriatrics Society, 54(11), pp.1641-1648.

[17]   Abras, C., Maloney-Krichmar, D., & Preece, J. (2004). User-centered design. Bainbridge, W. Encyclopedia of Human-Computer Interaction. Thousand Oaks: Sage Publications, 37(4), pp. 445-456.

[18]   Norman, D. A. (2002). The Design of Everyday Things, Basic Books, ISBN 0-465-06710-7.

[19]   Norman, D. A. (2005). Emotional Design. Basic Books. ISBN 0-465-05136-7.

[20]   de Ruyter B. & Pelgrim, E. (2007). Ambient assisted-living research in carelab. ACM interactions 14, 4 (July 2007), pp. 30-33. DOI=http://dx.doi.org/10.1145/1273961.1273981

V   Technical Concept of Health Data Collection and Integration
Data Analysis for Gaining Meaningful Medical Information

Serena Ponte / Elisa Ferrara / Silvana Dellepiane / Roberta Ferretti / Sonia Nardotto

Abstract

The collection of data for therapeutic purposes when using Serious Games in a home environment is essential to help therapists and medical doctors deliver better therapy. It is necessary to use different sensors for collecting such data. This requires using standards as HL7[1] as much as possible to provide a general purpose approach. On the other hand, the collected data needs an extensive evaluation and interpretation and tools are required to provide the therapist and medical doctor with meaningful information. In this chapter we offer a closer look to standards and propose tools for the analysis of collected data for later use by therapists and medical doctors as a result of the Rehab@Home project.

1	Introduction
2	Type of Data
3	Data Protocol
4	Rehabilitation Data Processing
5	Results
6	Conclusion
7	References

---

[1] HL7: Health Level 7 http://www.hl7.org/

© Springer Fachmedien Wiesbaden GmbH, part of Springer Nature 2018
M. Lawo und P. Knackfuß (Hrsg.), *Clinical Rehabilitation Experience Utilizing Serious Games*, Advanced Studies Mobile Research Center Bremen, https://doi.org/10.1007/978-3-658-21957-4_5

1   Introduction

The *Rehab@Home Operational Infrastructure*, as any such infrastructure of a similar kind, essentially relies on the acquisition, processing, exchange and interpretation of a large set of heterogeneous data and information coming from: existing clinical data records, rehabilitation workflow structures, user-system interactions, explicit user feedback, basic information about expected and actual rehabilitation progress, and biophysical sensors. This is a huge amount of raw input data to the system, which has limited value if the following minimum information relevant for the rehabilitation process is not properly extracted and managed:

- Initial rehabilitation profile and dynamic updating of progress;
- Biophysical status of the patient to understand his reaction to the proposed exercises.

The acquisition of many heterogeneous data and information related to the general status of a patient is done through the use of sensors and monitoring devices (see book chapters two and three). Then, tasks of data aggregation, selection, transformation and mining as well as the graphical presentation of the acquired data have to be addressed.

During the daily rehabilitation therapy, it is very important to understand the involvement and the emotions in order to give to the therapist the right information about the condition of the patient. Involvement and emotions can be expressed through many channels such as facial expressions, voice, but also physiological responses like accelerations and decelerations of the heart [1], [2]. For this reason, the heartbeat as a physiological signal is chosen to analyse its changes during all phases of the rehabilitation session. Patient's heartbeat, with a pulse oximeter transmitting data via Bluetooth, is recorded and, from it, the inter-beat is extracted.

After the storage of all sessions, data is analysed and processed with some particular algorithms that extract the most important features that, compared with data from game, give a lot of information about the progress and the involvement of the patient in the rehabilitation session. [3]

2 Type of Data

Data and signals of interest are of two types, according to their role and refer to local and remote use as explained in figure 1.

In the local section, the patient-user interacts with the game through a screen and the Rehab console with a sensor device like the Microsoft Kinect. At the same time, body sensors acquire and transmit the biophysical signals such as, for instance, the heart rate.

All this information is transmitted to the remote site, where the remote database receives the synthesis of the patient session from the game and the clinical patient data from the body sensors. The data client processes, aggregates, and appropriately summarizes all data.

The three classes of personal, clinical and gaming data define the data flow.

2.1 Personal Data

The information for Patient Identification (PD _ Identification of Patient) refers to personal data like name, gender, age, weight and height, and is collected as basic claims data when the patient enters the system.

At the first access on the database, this folder is created where the patient session archive of improvement or regression will be stored, as well as any other information of the patient, like historical biometric signals.

2.2 Gaming Data

The gaming data consist in the parameters describing the execution of the game. In particular, when referring to a movement and related spatial data, parameters of interest could be the number of repetitions, duration, speed, intensity, score over time, etc. as defined with the games in the previous book chapter.

Looking at figure 1, the gaming data flow starts at point (A), where the last level achieved in the patient game is loaded by the Rehab console for a correct restart of the game.

During the execution of any game a real-time feedback (B) in the order of milli- or microseconds allows the patient's interaction with the game, providing relevant feedback to the patient.

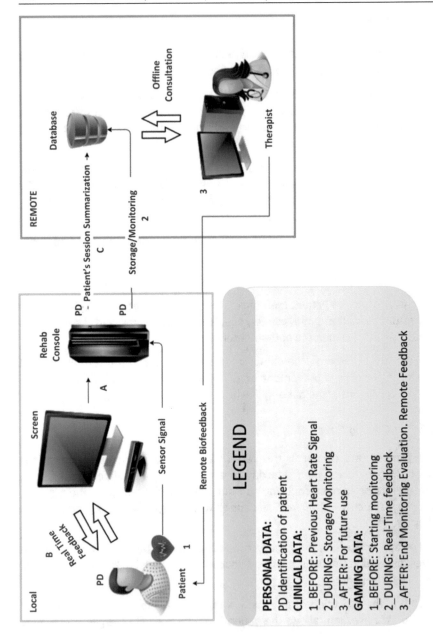

Figure 1 Data structure overview

At the highest level, or when the game ends (C), the results achieved and the eventual upgrade are transmitted to the remote database which gives the therapist a remote feedback for evaluation. Here a real-time response is not required. Storing the data and feedback can take up to a few minutes.

2.3   Biophysical Data

Biophysical data is acquired from the biophysical sensor applied to the patient. At point (1), the clinical data history is loaded, informing about previous problems or any abnormality. During the game, sensor data like the heart rate is acquired and stored (2). This stored data could be retrieved (3) from the database for any future consultation by the therapist or medical doctor.

3   Data Protocol

3.1   IHE ('Integrating the Healthcare Enterprise')

A major concern when dealing with data, signals, and information handling is integrability and unambiguous understanding of data. As is well known, a major worldwide objective in health care deals with the orderly exchange of information between diagnostic and therapy devices, clinical systems and IT systems from a variety of manufacturers within and across hospitals.

To this end, **IHE** ('Integrating the Healthcare Enterprise') [4] is an initiative by healthcare professionals and industry to improve the way computer systems in healthcare share information. IHE provides a common framework for building effective solutions to close the communication gaps between systems and foster their interoperability. High-quality patient care and optimized clinical workflows require efficient access to all relevant data across the continuum of care.

IHE promotes the coordinated use of established standards such as DICOM[2] and HL7 to address specific health needs in support of optimal patient care. When appropriate, the above guidelines for representation of data, signals, flows, and procedures, were adopted in the Rehab@Home project.

---

[2] DICOM: Digital Imaging and Communications in Medicine http://www.dicomstandard.org/

In all other cases, the huge internationally recognized work of standardization has been taken into account for a common and univocally defined data representation and integration protocol since it provides interesting indications and a basic model to describe the patient. This aspect assured a great level of integration suitability and created a common and unambiguous understanding.

In order to guarantee interoperability under the Rehab@Home operational infrastructure, suitable models were defined and implemented, to exchange heterogeneous data and information through their interfaces. The common data protocol aims at integrating the different data sources (health, physical, and personal, generated by the system, etc.) into a unique, properly designed structure.

The Signal Acquisition Component (SAC) as a modular platform allows the adding of several biomedical sensors attached to the patient's body. SAC carries out the raw acquisition and it does not process the data transmitted to the Signal Interpretation Component (SIC). The interoperable transmission of information was achieved through the exchange of instantiation of documents in appropriate format as the traditional XML, the easier and lighter JSON, or similar. This is explained in more detail in book chapter two.

To this end, a wide standardization work has been made in the biomedical and healthcare society, addressing various levels of information. A brief overview of the major standardization frameworks referring to integration, data and information exchange, and workflow procedures is given in the following.

In dealing with the specific exploitation of the acquired data, Rehab@Home system employed two main approaches, allowing the appropriate scalability of the signals involved: Whilst some signals and information are processed in real-time to provide immediate feedback to the patient during the rehabilitation session, some other data, signals, and information are to be stored in a central, remote database for subsequent consultation, reference, aggregation and processing purposes.

These two criteria address different aspects. In the former case the strict temporal constraint does not need always to allow the storage of the signals and data that are immediately used and processed. In such a case, no special standardization is required; the processing is driven by the specific hardware and software instrumentation as the related transmission protocol.

To this end, in the latter case, a precise and unambiguous information storage protocol is required following or inspired by existing standards.

We made a direct use of the commonly recognized standards such as HL7 and DICOM, exploited subsets of the defined entities and structures when feasible, and were inspired by these standards in all other cases, in order to define new and appropriate data representation. The wide world of grammatical, syntactical and semantical protocols already defined helped us to avoid the definition of new standards, maintaining the ability to also integrate with external systems.

3.2   HL7 (Health Level Seven International)

The Health Level Seven (HL7) [5] standard from the American College of Radiology (ACR) and the National Electrical Manufacturers' Association (NEMA) provides guidelines for data exchange to allow interoperability between healthcare information systems and focuses on the clinical and administrative data domains.

HL7 refers to the seventh layer of the ISO OSI reference model, also known as the application layer, independent of lower layers. The key goal is syntactic and semantic interoperability. HL7 standards define how information is packaged and communicated from one party to another, setting the language, structure and data types required for seamless integration between systems. HL7 standards support clinical practice and the management, delivery, and evaluation of health services, and are recognized as the most commonly used in the world.

HL7 standards [6], [7] are grouped into seven sections as reference categories:

- Section 1: Primary Standards - They address system integration, interoperability and compliance. The structure and semantics of 'clinical documents" are defined with the following six characteristics: 1) Persistence, 2) Stewardship, 3) Potential for authentication, 4) Context, 5) Wholeness and 6) Human readability.
- Section 2: Foundational Standards - They define the fundamental tools and building blocks as well as the technology infrastructure.
- Section 3: Clinical and Administrative Domains - This section defines messaging and document standards for clinical specialties and groups.
- Section 4: EHR Profiles - These standards provide functional models and profiles that enable the constructs for management of Electronic Health Records.

- Section 5: Implementation Guides - This section is for implementation guides and/or support documents.
- Section 6: Rules and References - Technical specifications, programming structures and guidelines for software and standards development.
- Section 7: Education & Awareness - HL7's Draft Standards for Trial Use (DSTUs), as well as helpful resources and tools to further supplement understanding and adoption of HL7 standards.

3.3   User Profiling & Personalization

Personalization involves a process of gathering user and process related information during the rehabilitation used to appropriately adapt process and services to satisfactorily enhance the user's experience. User satisfaction is the ultimate aim of personalization. Case-based reasoning will help in further improving the rehabilitation process by either exploiting profile information of the same patient or retrieving and processing information related to one or more similar patients.

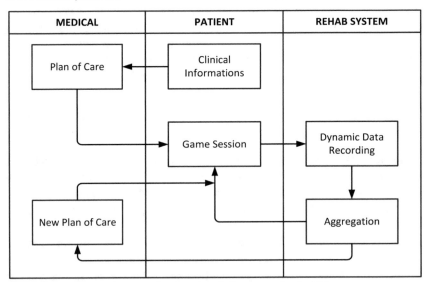

Figure 2 Procedure and personalization description according to IHE

Rehab@Home aimed to deal with the issue of rehabilitation from a 360° wide perspective. Specific and commonly shared standards for the description of user profile and rehabilitation procedures are mandatory. Only this ensures the design and development of a remote healthcare computational architecture based on distributed smart sensors (first processing layer) and personalized physiological assistive and intelligent algorithms integrated in a multi-level physiological human model (second processing layer). To this end, the IHE framework can be utilized or can suggest an appropriate procedure and personalization description (see figure 2).

As depicted in figure 2, the physician requires the patient's medical records and compiles a static care plan that allows the patient to make rehabilitation using the game. The patient receives a care plan to follow by using the Rehab console. After each game session, the patient's results are saved and compared with the expected improvement in order to check if the care plan had a positive effect on the patient. If the patient's results are satisfactory, the patient continues to play with the same care plan; if they are not satisfactory, the medical doctor needs to modify the care plan according to different needs of the patient.

4   Rehabilitation Data Processing

Data is processed and analysed with a specially implemented program that extracts the most significant features from the signal for each part of the rehabilitation session in which the protocol is divided.

4.1   Heartbeat Analysis

Heartbeat isclinical data easy to record and therefore easy to introduce in the home environment setting. This is the reason why a pulse oximeter was chosen as physiological sensor in the home-based rehabilitation system.[3]

The principal features extracted from the heartbeat are: average, standard deviation, maximum and minimum value and the difference between maximum and minimum value (see figure 4).

Dithering

The pulse oximeter has a quantization error making the signal too stationary over the time. Therefore we used a post processing technique called 'Dither-

---

[3] See book chapter three for technical details

ing'. This technique consists in adding a random noise with an appropriate distribution from -0.5 to +0.5 to our samples with the goal of minimisin distortion introduced by the truncation and thus obtaining a more realistic signal.

The two graphs of figure 3 show the effect when applying the algorithm to the heartbeat signal recorded during a game session.

It is clear that the dithered signal is less affected by the distortion introduced by the truncation if the data were quantized again.

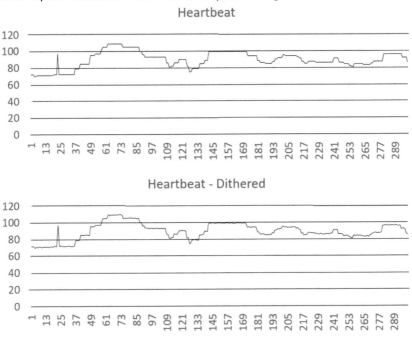

Figure 3 A section of dithered data

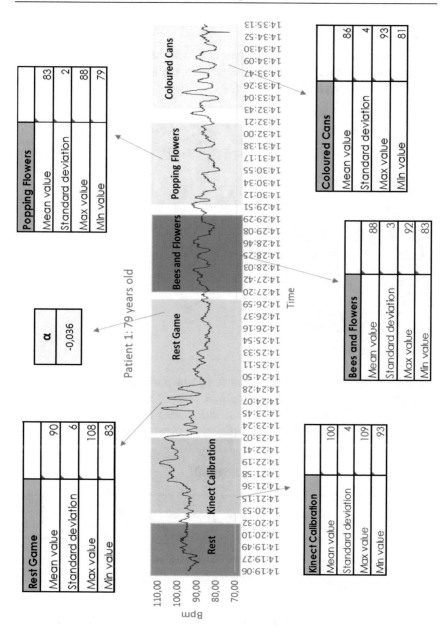

Figure 4 Example of heartbeat signal recorded during a rehabilitation session. Each coloured window represents a different game phase.

4.2   Score Analysis

From the game session, a sort of report file, with the main information from the session, is generated, such as final score, score over time, spatial feedback and right and wrong items (see figure 5)

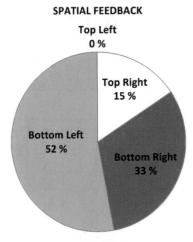

Figure 5 Spatial feedback

Once games data is collected, the main game parameters are extracted and processed in order to give a clear graphical view of them. Many types of graphs were created in search of the best to summarize the results. The graphs in figure 6 and figure 7 give some examples of this processing.

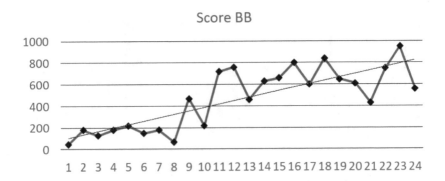

Figure 6 Graphical example of score over time

These graphical portrayals could be helpful for the therapist to evaluate and supervise the patient's progress during the therapy. For example, as it can be seen in figure 7, the patient has improved in some games while in others, some difficulties can be observed.

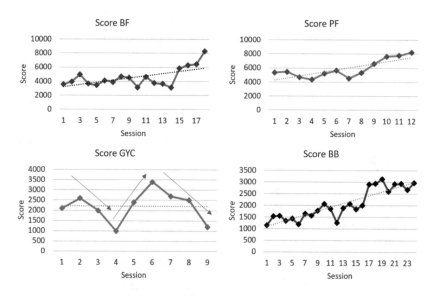

Figure 7 Score example collected by a patient during three rehabilitation sessions

During our trials with patients, the performance of games related to rehabilitation sessions was analysed in order to evaluate any improvement. The scores' data was extracted from Rehab@Home database. For each patient, we analysed the scores of 4 different games (see previous book chapter). In several sessions the patient, according to the instructions of physiotherapists, could run games for 60, 120 or 180 seconds. When transcending the game duration, which would allow the patients to acquire a higher score, the data was evaluated only for the first 60 seconds or 120 seconds (see figure 8).

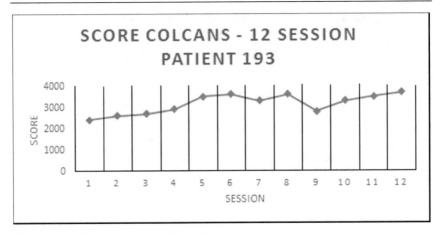

Figure 8 Score Curve of Coloured Cans Game

Subsequently, in order to have an overall picture of the progress achieved during the rehabilitation program, we also evaluated the overall scores of 4 games achieved during the 12 sessions (see figure 9).

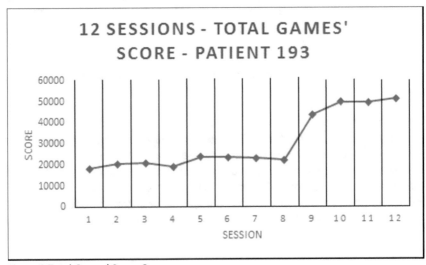

Figure 9 Total Games' Score Curve

4.3   Heart Beat and Score Trend Analysis

If the patients' state is really improved we should expect that, with the sessions progress, they could acquire more confidence with games, improving the mobility of their arts, being less stressed and stirred, and struggling less than during the first session. This translated in terms of data analysis would indicate an increasing trend of the score curve and a decreasing trend of the heartbeat curve.

Heartbeat, recorded during the execution of each game, gave us 120 beats per minute, so before performing the score normalization, the heartbeat mean is evaluated by equation (1) in order to have a one-to-one correspondence between the two signals:

$$S' = \frac{S_i - S_{min}}{(\frac{S_{max} - S_{min}}{B_{max} - B_{min}})} + B_{min} , \tag{1}$$

where

S_i is each score point I,

S_{Min} is the minimum of all scores points,

S_{Max} is the maximum of all scores points,

B_{Min} is the minimum of all beats per minute points,

B_{Max} is the maximum of all beats per minute points,

S' is the score points normalized.

Linear Regression

Since the trend is common information used in data evaluation and since it is extractable from both signals, it was chosen to calculate the angular coefficient of the regression lines in order to evaluate the presence of a correlation between the heart beat and the score. Linear regression is the most commonly used technique for determining how one variable of interest (the response variable) is affected by changes in another variable (the explanatory variable). The variable we are predicting is called the *criterion variable* and is referred to as Y. The variable we base our predictions on is called the *predictor variable* and is referred to as X. When there is only one predictor variable, the prediction method is called *simple regression*. In simple linear regression, the topic of this section, the predictions of Y when plotted as a function of X form a straight line (see figure 10) [8].

The linear model is expressed by equation (2):

$$Y = \beta_0 + \beta_1 X + \varepsilon,\tag{2}$$

where

Y denotes the dependent variable,

X denotes the independent variable,

β_0 denotes the Y-intercept,

β_1 denotes the slope,

ε denotes the error variable.

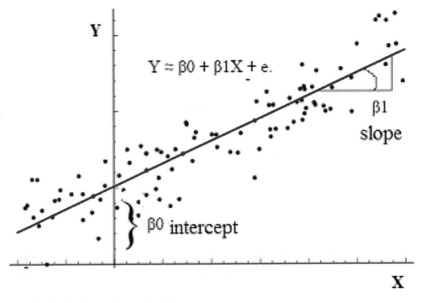

Figure 10 Simple Linear Regression Line

The best fitting line is the one that minimizes the least square fit:

$$Q = \sum_{i=1}^{N}(Y_i - (\beta_0 + \beta_1 X_i))^2 = \sum_{i=1}^{N}\varepsilon_i^2.\tag{3}$$

Calculating the partial derivatives and set them equal to zero we have

$$\frac{\partial Q}{\partial \beta_0} = -2 \sum (Y_i - (\beta_0 + \beta_1 X_i)) = 0 \tag{4}$$

and

$$\frac{\partial Q}{\partial \beta_1} = -2 \sum X_i \left(Y_i - (\beta_0 + \beta_1 X_i) \right) = 0. \tag{5}$$

After a few algebraic calculations we find the coefficients $\overline{\beta_0}$ and $\overline{\beta_1}$ expressed as

$$\overline{\beta_1} = \frac{cov(X,Y)}{S_i^2} \tag{6}$$

and

$$\overline{\beta_0} = \overline{Y} - \overline{\beta_1} X \tag{7}$$

where:

$cov(X, Y)$ is the covariance, a measure of the strength of the correlation between two or more sets of random variables and.

S_i^2 is the correlation coefficient as quantity giving the quality of a least square coefficient to the original data.

The regression equation that estimates the linear model is

$$\tilde{Y} = \overline{\beta_0} - \overline{\beta_1} X . \tag{8}$$

In order to analyse the data recorded during the sessions, score and heart-beat's slopes are calculated. These two parameters are represented as the coordinates of a dispersion diagram in order to evaluate their influence on the session trend. In figure 11 we give an example of the linear regression associated to a patient's data recorded during the performance of four games in 12 sessions.

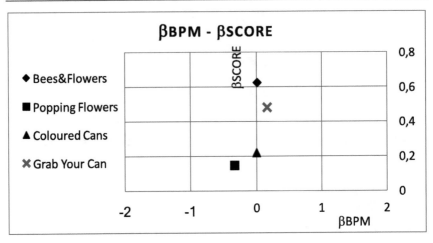

Figure 11 Linear Regression - example of Patient 186

Four cases can be distinguished:

(1) $\beta_{bpm} > 0$ and $\beta_{score} > 0$

Heartbeat increasing and score increasing, therefore stressed patient but with improvement in the rehabilitative performance;

(2) $\beta_{bpm} > 0$ and $\beta_{score} < 0$

Heartbeat increasing and score decreasing, therefore stressed patient but with worsening of the rehabilitative performance;

(3) $\beta_{bpm} < 0$ and $\beta_{score} < 0$

Heartbeat decreasing and score decreasing, therefore relaxed patient but with worsening in the rehabilitative performance;

(4) $\beta_{bpm} < 0$ and $\beta_{score} > 0$

Heartbeat decreasing and score increasing, therefore relaxed patient but with improvement in the rehabilitative performance, so the ideal situation.

4.4   Inter Exercise Time

Another parameter which has been evaluated is the rest time between the end of one exercise and the beginning of the following one. This parameter is obtained from the combination of the game log and the duration.

The inter exercise or rest time is calculated as follows:

$$InterTime_{n-(n+1)} = LEx_{n+1} - (LEx_n + Duration_n), \qquad (9)$$

where

LEx_n is the beginning of an exercise,

$Duration_n$ is the duration of the exercise and

LEx_{n+1} is the beginning of the following exercise.

The result for all sessions is represented as bar plot for each patient, as seen in figure 12.

This parameter is important for the therapists because it allows them to figure out if the patient is too fatigued while performing an exercise or needing to rest before performing the next one.

To deepen the analysis, the inter-exercise time average, standard deviation, maximum and minimum have been extracted. The twelve sessions carried out by the patients were divided into two groups: the first six and last six to evaluate whether the efficacy of the therapy was going to affect also the length of breaks and then the execution flow of the exercise sessions.

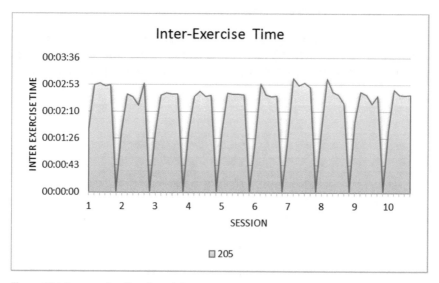

Figure 12 Inter exercise time bar plot

Finally, the inter exercise time average of the first and the last six sessions was calculated to evaluate the average improvement of the subjects.

4.5   The Rest Phase

Also for the rest phase, a linear regression is computed to extract parameters helping us to understand the feelings, the physiological condition of the patient, and general reactions to the rehabilitation exercise.

On the rest phase window, the straight line that best approximates the signal is computed and, from it, the angular coefficient α is extracted.

From the game data, some conclusions can be extracted:

- If the score over time at the beginning of the session is zero or very low, it is possible that the user has no clear aim for the game, or the level proposed is too difficult.
- If the score over time at the end of the session is zero or very low, it could mean that the user is tired and fatigued; so next time the duration of the game should be set shorter by the therapist.
- If the patient scores always lower than the maximum score achievable, the therapist should extend the duration.

Key points extracted both from heartbeat and game can be compared and analysed in parallel. This will allow us to extract some type of correlation (if present) to understand the meaning of the relation between the two signals.

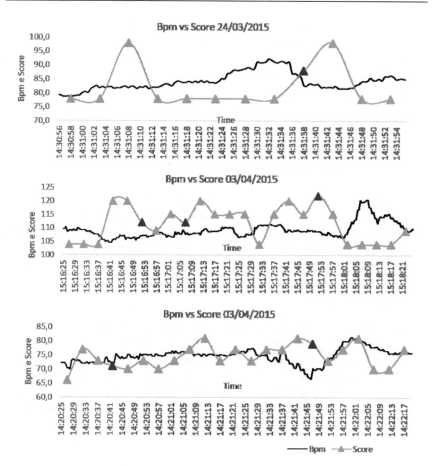

Figure 13 The black line represents the heartbeat signal, while the triangles represent the score. Dark indicators appear when the user makes a mistake during the game.

Cross correlation is a standard method of estimating the degree to which two series are correlated. Consider two series x(i) and y(i) where i=0,1,2...N-1. The cross correlation 'r" at delay d is defined as

$$ r = \frac{\sum_{i} [(x(i) - mx) * (y(i-d) - my)]}{\sqrt{\sum_{i} (x(i) - mx)^2} \sqrt{\sum_{i} (y(i-d) - my)^2}} \quad (10) $$

where m_x and m_y are the means of the corresponding series.

In figure 13, the black line represents the heartbeat signal, whilst the line with triangles shows the score. Dark indicators appear when the user makes a mistake during the game. Figure 13 is an example of combined score and heartbeat graphs as a possible representation of data acquired and processed after the rehabilitation session.

5   Results

5.1   Signal Acquisition Component

The Signal Acquisition Component (SAC) is a modular platform that allows the adding of several biomedical sensors attached to the patient's body. SAC carries out the raw data acquisition and it does not elaborate the data transmitted to the Signal Interpretation Component (SIC) for this purpose.

The selection and integration of many different components coming from a heterogeneous set of technological fields is a requirement for monitoring the rehabilitation progress.

Nowadays, many different kinds of sensors are commercially available, suitable for measuring both environmental and physical data like position, location, acceleration, force, direction, weight, pressure. Even cameras for pictures or video acquisition are available. Physiological data such as temperature, spirometry, saturation, blood pressure, blood glucose, ECG[4] and EEG[5] provide valuable parameters. The idea is to put different kinds of sensors in the patient surroundings in order to monitor the execution of rehabilitation exercises using Serious Games.

5.2   Signal Interpretation Component

The Signal Interpretation Component (SIC) is processing software with the aim of monitoring the level of interest, effectiveness and difficulty perceived by the patient during the game.

The SIC receives a critical threshold like the upper limit for heart rate, set by therapist in the medical chart. It continuously checks if the patient is about to cross that limit comparing it with the data stream from SAC. A built in algo-

---

[4] ECG: Electrocardiography
[5] EEG: Electroencephalography

rithm provides an alert in case the user should stop the game. All biomedical data sent by SAC to SIC during a session is temporarilyy logged.

5.3   Data Client

The Data Client component is part of the back-end structure of the Reference Rehabilitation Platform as described in chapter two. It retrieves data from the Central Information Space (CIS) storing the raw data received from Game Component (GC) and SIC for each game session. Data is processed by algorithms whose task is to highlight the problems during the game like moments of high excitement or else to understand if the game intensity is too low or high for the patient. To perform this kind of computation, the Data Client is linked to a Relational database management system (RDBMS) where it stores data useful to achieve a data mining process finalized to introduce a certain amount of intelligence in data analysis. The quality and precision of the Data Client output is supposed to improve as soon as a sufficient amount of data coming from the game sessions is available for exploitation. The Data Client output is called summarization and it is stored in the CIS with the right format as required by Therapist Client interface.

5.4   Rehab@Home Database

The Data Client relies on a relational database management system (RDBMS) to improve its summarization analysis. Storing data in a SQL[6] database allows an easy access to the information for any developer that could like to access directly to them, simply running MySQL queries in order to filter or compare some results.

Each patient is identified by an ID that maintains his personal data anonymously. In figure 14 we give a screenshot of the Rehab@Home database.

---

[6] SQL: Structured Query Language - https://www.w3schools.com/sql/

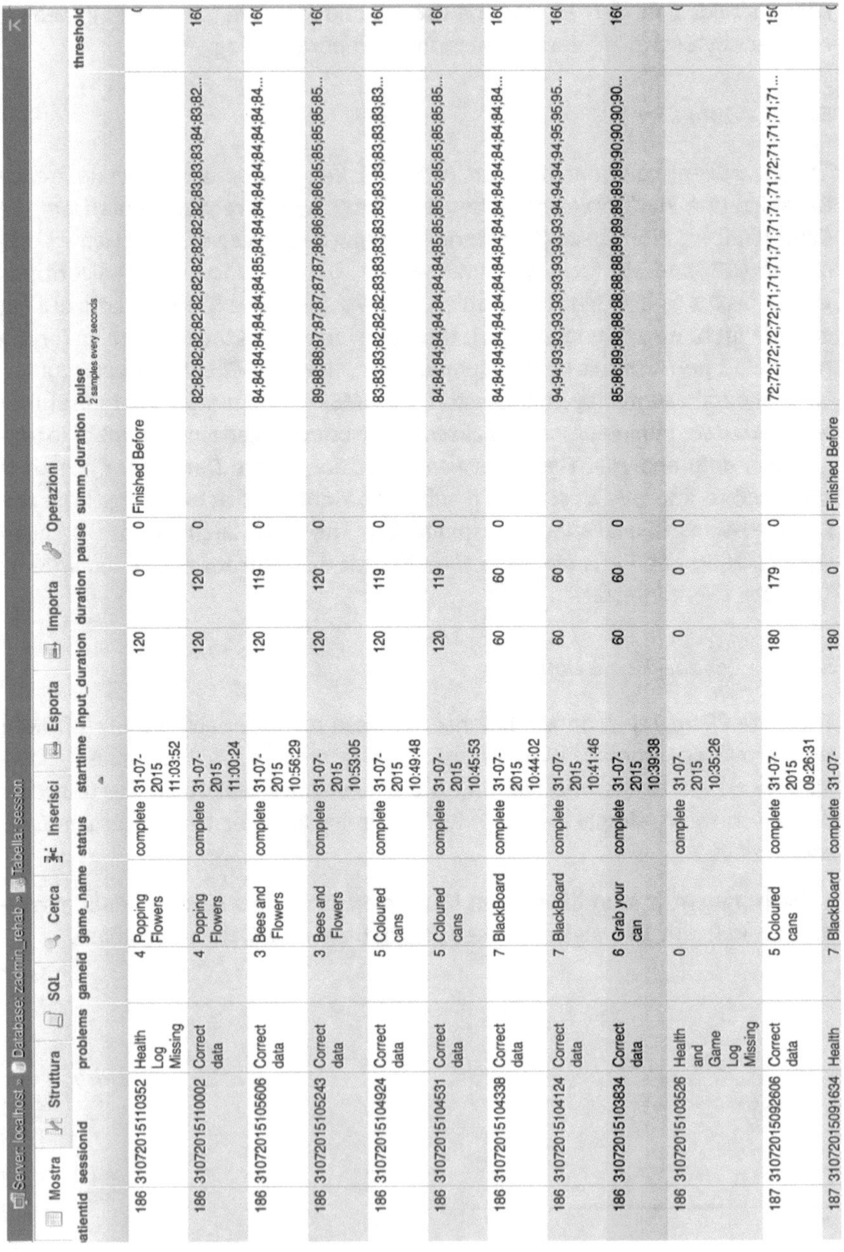

Server: localhost » Database: zadmin_rehab » Tabella: session

Mostra | Struttura | SQL | Cerca | Inserisci | Esporta | Importa | Operazioni

atientid	sessionid	problems	gameid	game_name	status	starttime	input_duration	duration	pause	summ_duration	pulse 2 samples every seconds	threshold
186	31072015110352	Health Log Missing	4	Popping Flowers	complete	31-07-2015 11:03:52	120	0	0	Finished Before		0
186	31072015110002	Correct data	4	Popping Flowers	complete	31-07-2015 11:00:24	120	120	0		82;82;82;82;82;82;82;82;82;82;82;82;82;83;83;83;83;84;83;82...	16
186	31072015105606	Correct data	3	Bees and Flowers	complete	31-07-2015 10:56:29	120	119	0		84;84;84;84;84;84;84;85;84;84;84;84;84;84;84;84;84;84...	16
186	31072015105243	Correct data	3	Bees and Flowers	complete	31-07-2015 10:53:05	120	120	0		89;88;88;87;87;87;87;86;86;86;85;85;85;85;85...	16
186	31072015104924	Correct data	5	Coloured cans	complete	31-07-2015 10:49:48	120	119	0		83;83;83;82;82;82;83;83;83;83;83;83;83;83;83;83...	16
186	31072015104531	Correct data	5	Coloured cans	complete	31-07-2015 10:45:53	120	119	0		84;84;84;84;84;84;84;85;85;85;85;85;85;85;85;85...	16
186	31072015104338	Correct data	7	BlackBoard	complete	31-07-2015 10:44:02	60	60	0		84;84;84;84;84;84;84;84;84;84;84;84;84;84;84;84...	16
186	31072015104124	Correct data	7	BlackBoard	complete	31-07-2015 10:41:46	60	60	0		94;94;93;93;93;93;93;93;94;94;94;94;95;95;95...	16
186	31072015103834	Correct data	6	Grab your can	complete	31-07-2015 10:39:38	60	60	0		85;86;89;88;88;88;88;89;89;89;89;90;90;90;90...	16
186	31072015103526	Health and Game Log Missing	0		complete	31-07-2015 10:35:26	0	0	0			0
187	31072015092606	Correct data	5	Coloured cans	complete	31-07-2015 09:26:31	180	179	0		72;72;72;72;72;71;71;71;71;71;71;71;72;71;71;71;71...	15
187	31072015091634	Health	7	BlackBoard	complete	31-07-	180	0	0	Finished Before		0

Figure 14 Screenshot of data collected in the Rehab@Home database

6 Conclusion

Serious Games offer, in a new way, the opportunity to collect clinical data related to therapy involving movement. A large amount of data becomes available. Using this data enables a much deeper study of effects like the correlation of gaming and clinical data. Using standards where ever possible allows the integration of different games as medical devices. The therapist receives a much better basis for helping the patient, although Serious Games do not seek to provide the therapy itself.

7 References

[1]   Nacke, L. E., Kalyn, M., Lough, C., & Mandryk, R. L. (2011, May). Biofeedback game design: using direct and indirect physiological control to enhance game interaction. In Proceedings of the SIGCHI conference on human factors in computing systems, pp. 103-112. ACM.

[2]   Nacke, L. E., & Lindley, C. A. (2010). Affective ludology, flow and immersion in a first-person shooter: Measurement of player experience. arXiv preprintarXiv:1004.0248.

[3]   Nacke, L. (2009). Affective ludology: Scientific measurement of user experience in interactive entertainment.

[4]   IHE Patient Care Coordination Technical Framework Supplement, Cross Enterprise TeleHomeMonitoring Workflow Definition Profile (XTHM-WD). Date: August 16, 2012. Author: IHE PCC Technical Committee

[5]   http://www.hl7.org, accessed 9.8.2017

[6]   http://www.hl7.org/implement/standards/, accessed 9.8.2017

[7]   http://www.hl7.org/implement/standards/fhir/, accessed 9.8.2017

[8]   David A. Freedman (2009). Statistical Models: Theory and Practice. Cambridge University Press.

VI Media Ecology Aspects of Homecare Assistive & Rehabilitation Technology How to Integrate into the User Centred Design Process

Dov Faust / Edna Pasher / Robert K. Logan / Hadas Raz

Abstract

Modern societies are facing two major trends: widespread population ageing and rapid development of new technologies. Since old age is usually also a time of reduced and diminished abilities and health, it is very important to recognize the potential of technological advances to enhance health, abilities and relationships. However, the abilities, needs, aspirations and contextual environments of older people vary greatly. This chapter gives an overview of the characteristics and circumstances of the older adults that are perceived to have potential influence on the acceptability of assistive and rehabilitation technology systems and devices. We explore the concept of 'Media Ecology' and its impact on the adoption success of assistive and rehabilitation technology systems and devices by elderly population.

1	Introduction
2	The Need
3	Homecare A&RT Media Ecology – Challenges and Opportunities
4	Overcoming A&RT Media Ecology Barriers: Designing for the Elderly – User Centred Design (UCD)
5	Conclusion
6	References

© Springer Fachmedien Wiesbaden GmbH, part of Springer Nature 2018
M. Lawo und P. Knackfuß (Hrsg.), *Clinical Rehabilitation Experience Utilizing Serious Games*, Advanced Studies Mobile Research Center Bremen, https://doi.org/10.1007/978-3-658-21957-4_6

1   Introduction

Home Assistive and Rehabilitation Technology (A&RT) seeks to narrow the gap between an individual's capacity and their living environment and therefore to make it easier for people to remain in their existing accommodation [1]. The extent to which Assistive Technology (AT) can narrow this gap depends on older people's acceptance and willingness to use it which depends on several complex factors to be discussed. There are numerous innovations in computer technological applications aimed at helping older people to retain independence and social engagement while overcoming disabilities caused by either the natural ageing process or by acute events requiring rehabilitation. Technology can make a major contribution to helping older people in various wellbeing aspects through living aids, wearable devices, exercise apps, gaming and e-books for mental stimulation or social networking for companionship [2]. Although different approaches and concepts have been used, historically, personal computer based systems usage by the elderly has been relatively low and some have attributed it to lack of interest on the part of the older people.

Incorporation of interactive computerized assistive technology into the healthcare services has a dramatic influence on the whole ageing population eco-system in respect to health improvement outcomes, hospitalization reduction and cost effectiveness of the whole process [2]. However, there is a need to well understand and resolve the impact of various barriers and drivers of health A&RT use, from personal, social and organizational aspects.

The idea behind Media Ecology [3] is that, regardless of its content, a medium itself has certain effects on the way in which the user interacts with that medium. This view is contained in McLuhan's signature one-liner 'the medium is the message'. We claim that the relationship between the system design process and the holistic spectrum of human, social and organizational factors is the key to A&RT systems and devices acceptance by the ageing, chronically ill and disabled population.

2   The Need

2.1   Ageing Population Demographic Trends

According to the UN Department of Economic and Social Affairs [4], significant population ageing is taking place worldwide (see figure 1). Ageing results from both decreasing mortality and declining fertility. This process leads to a pro-nounced increase of the older persons' share in the population. The global share of older people (aged 60 years or over) increased from 9.2 % in 1990 to 11.7 % in 2013 and will continue to grow as a proportion of the world popula-tion, reaching 21.1 % by 2050.The older population is itself ageing. Globally, the share of older persons aged 80 years or over (the 'oldest old') within the older population was 14 % in 2013 and is projected to reach 19 % in 2050. If this projection is realized, there will be 392 million persons aged 80 years or over by 2050, more than three times the present figure.

This trend is creating fiscal pressures on home healthcare support and reha-bilitation systems for older persons. In a number of developing countries, pov-erty is high among older persons, sometimes higher than that of the popula-tion as a whole, especially in countries with limited coverage of social security systems. Older persons can increasingly live independently (alone or with their spouse only) but, in most countries though, they have to support themselves financially. While people are living longer lives almost everywhere, the preva-lence of non-communicable diseases and disability increase with population age.

Although nowadays people are living longer, the increase in life expectancy is accompanied by an increase in the number of years lived with disabilities. It can be expected, as the population and life expectancy of older people is rising, that the number of those living with some kind of disability will increase in the future.

As society ages, older people's living arrangements will undoubtedly be-come a major component within mainstream housing. Heller et al. [5] say about aging: 'As people grow older, their abilities change. This change includes a decline in cognitive, physical and sensory functions, each of which will decline at different rates relative to one another for each individual.' This makes it hard to define 'the elderly' as one consistent group and presents a challenge for assistive technology systems designers. Facing this challenge, however, can ultimately provide great benefits for both elderly people and the society.

Unfortunately, in our modern society, the way of caring for the elderly is economically unsustainable because it is based on a costly, hospital-centred health system. Cost for the hospitalization and healthcare services of the older population is enormous and fast growing, taking into account the costs for emergency treatment, medical professionals, drugs, and physician's services.

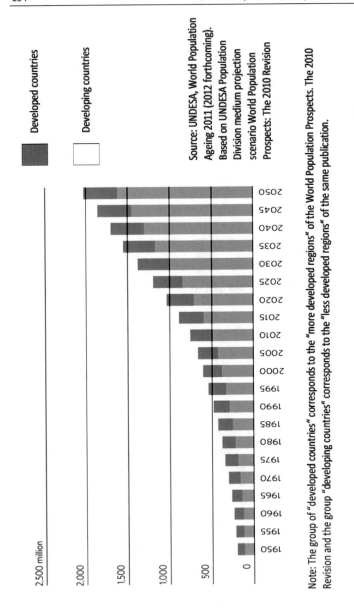

Figure 1 Number of people aged 60 or over: World, developed and developing countries, 1959-2050

Besides the big burden for the families involved, it is also a challenge for the community to take care of those persons or to reconstruct homes for the special disabilities needs [2].

The increasing costs for both routine and rehabilitation services lead to a reduction in the recovery period in the medical facilities, less personal attention and often releasing patients too early. This has a negative impact on the quality of life of patient and families.

If we are to advise a better alternative, we need to start by understanding the real needs of the elderly and why we currently spend so much on their care. In this context, suitable A&RT aids at home can help to prolong the medical treatment together with an adequate exercise from a personalized training programme instead of driving the patient to and from hospital and healthcare facilities. Technology can make a major contribution in helping older people to maintain their health and independence by leveraging mobile technology, Internet of Things (IoT), detection sensors software, wearable devices, exercise applications, gaming and e-books for mental stimulation or social networking for companionship.

Computer based AT may help older people to:

- Stay independent and healthy for as long as possible;
- Manage simple chronic conditions;
- Remain independent, overcoming complex co-morbidities;
- Minimize the time they have to spend in hospital;
- Find the right residential care when they require it.

However, in order to exploit the potential of the technological innovative solutions, one has to carefully look at the whole healthcare eco system or in other words the Media Ecology of A&RT, deeply examining the human, social and organizational aspects involved in the usage of it and to overcome serious barriers related to the interface between the users, their needs and the environment they are living in.

Two developments need to be taken into account: Aging of the population (see figure 2) and the increase of dementia (see figure 3) as a result out of this. The United Nations predict there will be more pensioners than children by 2050. That health conditions like dementia will continue to rise is a kind of time bomb.

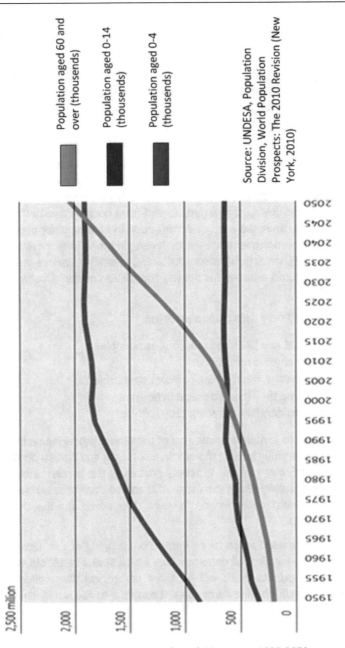

Figure 2 Population aged 0-4, 0-14 and aged 60 or more 1950-2050

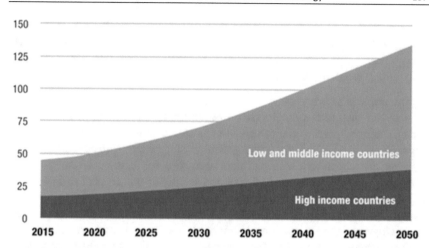

Source: World Health Organization and Alzheimer's Disease International Dementia: A Public Health Priority (Geneva, 2012)

Figure 3 Growth in numbers of people with dementia in high-income and low-and middle-income countries

2.2   Design Thinking – Problem Finding, Framing and Solving

The mere formulation of a problem is far often more essential than its solution, which may be merely a matter of mathematical or experimental skill. To raise new questions, new possibilities, and to regard old problems from a new angle requires creative imagination and marks real advances in science [24] and design.

Design involves problem finding, problem solving, analysis, invention and evaluation guided by a deep sensitivity to environmental concerns and human-centred aesthetic, cultural and functional needs [25]. - Charles Owen (2004) Design thinking is a holistic systemic approach to design that takes into account the full gamut of end users' needs as well as the needs of the innovators creating a new product, service or business. It consists of the following three elements:

- Problem finding or opportunity analysis as we outlined above;
- Problem framing through design research;
- Problem solving or service development through traditional design activities.

Scott Berkun in his book '*The Myths of Innovation*' explodes a number of myths including the one that the key to innovation is problem solving [26]. He argues that, while problem solving is a part of the innovation process, problem finding and problem framing are actually more important.

The order in which these three activities are pursued is essential to the success of the overall design process. They should be performed in the order in which they are listed. Too often companies and product developers do it in the opposite direction. The first two activities of problem finding and problem framing are absolutely essential and necessary for innovative design work. They must come first before the problem solving or the product development aspect of design can start. Too often problem finding and problem framing are bypassed and designers or technology developers start immediately with their technology and the product development phase. This can lead, and often does, to their solving the wrong problem. In cases where they were lucky enough to have picked the right problem to solve they can, and often do, fail to take into account the users' needs, expectations, desires or latent wishes or they fail to be guided by the demands and constraints of the market place and/or those charged with manufacturing and delivering the product or service.

Problem finding entails opportunity analysis. We must first determine what emerging technologies and changing patterns of human behaviour will lead to the new products and services of the future. Next we must consider what form these new products or services will take and how we will be able to imagine them. We believe that imagining innovative products, services and businesses requires strategic foresight. According to Hines and Bishop [27] strategic foresight helps one to make better, more informed decisions in the present – making it the ideal tool for exploring new products, and services for the elderly.

Problem framing requires design research and design thinking. We might have an idea for an innovation to address a need, but is it commercially viable and is it technically feasible? What user needs research is necessary to be conducted to help the elderly? Are our solutions user-centred and will there be a demand for them? The design research that is carried out to frame the problem must be grounded in ethnographic techniques of observation of the potential users and their sense of the potential value proposition of the services we intend to bring to the elderly. The dynamics of the market place and the socio-cultural milieu in which the innovations we are proposing will operate must be understood. One must answer the question of what unmet needs will be addressed by the planned innovation in terms of tangible outcomes and benefits for the elderly. Finally, the framing exercise should lead to parameters for creating prototype services or offerings that can be empirically tested with potential users.

Problem solving requires the traditional design activities of aesthetics, usability, branding, marketing and the generation of a business model. This final step is critical to the final success of the product, service or business.

In conclusion, the development of the innovations we are proposing should not be undertaken until the problem finding and framing process is complete or well underway. It is possible that once the problem finding phase is basically complete and progress has been made in the problem framing phase some advantage could be gained by starting the development phase with the understanding that one should then toggle back and forth between the problem framing and the problem solving activities until a satisfactory solution is found. The development of the innovative services we are proposing is both an emergent and an evolutionary process.

3   Homecare A&RT Media Ecology - Challenges and Opportunities

When we come to try and understand the reasons for the phenomena of elderly people avoiding usage of computerized A&RT we have to look at the 'Big Picture'- the whole eco-system: personal physical and psychological reasons, social reasons and healthcare providers and infrastructure reasons.

3.1   About Media Ecology and Homecare A&RT

Following the ideas of J. C. R. Licklider, Andy Clark, Marshall McLuhan, Neil Postman and the concept of Media Ecology, the interface between users and computers technology in general and with assistive technology in our case, requires understanding the environment created by both the computerized system and its impact on individual (physical mental and psychological), social and organizational aspects [3]. As McLuhan pointed out, one cannot understand a figure unless one understands the environment in which it operates. This concept of Media Ecology applies to computerized assistive technology, wearable devices rehabilitation aids and to any other human and advanced technology iterative interaction (see figure 4).

Human factors can be divided to human attributes such as user qualification, knowledge and skills and to factors that are the result of the human-machine interaction such as empowerment, privacy etc.

Figure 4 Media Ecology Factors Model [17]

Social factors refer to the interaction of a user with other people, either a one-to-one interaction or an interaction with a group. For example, the impact of wearable computer on social identity.

Organizational factors are the impacts of assistive technologies on the organizational environment in terms of culture, structure and processes, work relations customer-supplier relations etc. [17]

These factors overlap, are interdependent and influence each other. Some might be conflicting; their influence changes based on the specific environment and circumstances.

3.2 Homecare A&RT Media Ecology - General Aspects

Whilst the ageing process is different for everyone, we all go through some fundamental changes. Not all of them are exactly what we expect. Ageing makes some things harder and one of those things is using technology.

There are numerous assistive technologies available but, unfortunately, it is well known that they often remain unused [6].

Many studies and researches have been taken place over the years pointing out many factors, categories and aspects related to the older person and the environment that may potentially influence the acceptability of computerized A&RT systems/devices among them are:

- Nature and degree of the personal limitations older adults might be facing in terms of vision, hearing, mobility and cognition;
- Psychological factors like attitudes to ICT, lack of interest and motivation to use ICTs, lack of confidence and fear of the ICTs, willingness to learn and acquire ICT-related skills, privacy issues, recognition of ICT quality and its usefulness, and expectations, needs and interests of the older person;
- Socio-demographic factors like age, gender, income, etc.

Richardson et al. identified three major categories of barriers that impeded the use and learning of computer technology for older people [28]:

- Person-centred barriers, such as emotional, mental and physical states;
- Learning environment barriers, such as a lack of emotional and practical support from others and age-unfriendly tuition;
- Individual circumstance barriers, such as a lack of a perceived need for a computer or the ability to buy one and to maintain computer hardware and software.

In investigating factors that might prevent older people from enjoying the benefits of computer technology or technological devices in general, researchers have identified barriers to use that can be categorized into five areas:

- Financial issues, including the high cost of computer acquisition and tuition for people on pensions;
- Learning and training barriers;
- Lack of motivation, including a perceived lack of need for or interest in computers and computing

- Physical and cognitive problems in using a computer keyboard and mouse due to arthritis and other health problems; and
- Lack of social support due to the absence of friends and relatives encouraging them to learn.

The above have to be linked to organizational barriers and limitations of infrastructure, practices and work processes, supporting needs for the full transformation into the digital homecare era. Below we elaborate on the various elderly users' acceptance and rejection factors along the three A&RT Media Ecology dimensions: personal, social and organizational.

3.3   Homecare A&RT Media Ecology - Human Aspects

Older adults might be facing physical limitations different in nature and degree either as part of the natural gradual ageing process or due to a health event like a stroke in terms of vision, hearing, mobility and cognition [7]: The decrease of vision [8] can mean the reduction in width of visual field, light sensitivity, colour perceptions, resistance to glare, dynamic and static acuity, contrast sensitivity, visual search and processing, and pattern recognition. The decrease in hearing goes mainly along with difficulties in hearing high frequency sounds. The decrease of motor skills [9] can lead to a decreased speed of movement, decline in strength and endurance, changes in balance and coordination, involuntary movement, tremor, restlessness, flexed posture. These changes have dramatic impact on movements, reaction time and accuracy. Cognitive changes [10] mean cognitive skills deterioration, affecting learning capability. Age related memory changes and their effects on learning are the main reason for the difficulties older people have in using computers. As Zajicek points out [10], other research shows that older people have more difficulties in retracing and navigating a route, which can be compared with the type of navigational skills needed on the Internet.

Whether a specific A&RT system will be used or not depends, amongst other reasons, on users' perceptions of themselves. The social model of disability recognizes that people may not define themselves as disabled or in need of special equipment so they may not take up a service offered with the best of intentions. Other possible reasons for rejection of assistive technology use are: Technology was incorrectly prescribed or imposes too great a burden in use. Often, however, it is the stigma attached to an assistive product. Services provide technology to meet a 'need', but users (and their care providers) will most readily use technology that is desirable because it enhances their social status as well enabling them to do things or making them feel better.

Among the main factors affecting acceptance and rejection of technical aids are the fear of the new, lack of motivation for use; often demand is lacking for this specific function or people are unwilling to try it out. In addition, it can be the ease or complexity of use or the lack of advice, training and encouragement.

Training and ongoing support helps older people overcome some of their anxieties, build skills and develop their confidence in using technology. The view of most experts is that we have all the kit that is needed. What we lack is the human element: the people and programs to deliver the necessary training and support .

'Diffusion Theory' [11] points out that whether an individual adopts a technological innovation such as computers depends on an interplay between contextual characteristics of the individual and socioeconomic status, health, beliefs about the technology like its complexity, and the perception of need for the technology, e.g. how will a computer help me?

Saying all this, it is important to mention that the heterogeneity of older people and the diversity of their living circumstances means that individual preferences will play a strong part in people's attitudes and assistive systems adoption accordingly.

3.4   Homecare A&RT Media Ecology - Social Aspects

Elderly social environment is sometimes underrated in this area [12] of using A&RT as most attention is given to personal cognitive, physical and sensory aspects. It has, however, become clear that the use of technology by elderly people has a direct influence on their social environment and healthcare outcome. Computer technology has a significant influence on the social life of elderly people, playing an important role directly and indirectly on their quality of life, including healthcare and rehabilitation services.

Effective computerized A&RT systems require an on-line data sharing network, enabling close monitoring and tracking of critical health parameters. Older people have different social relationships to young people. Research studies on how older people interact with healthcare professionals show that, in many cases, they have seen the same doctors for decades, leading to a very high degree of trust. But due to health and mobility issues, the 'world' available to the elderly is often smaller both physically and socially. Digital technology has an obvious role to play here by connecting people virtually when being in the same room is hard.

As written by Karavidas et al. [13] 'computers can present unique opportunities for older adults to socialize and establish social networks that can help alleviate loneliness and alienation.' For older people with mobility problems or older people that live geographically away from their family members and/or their care providers, computer-based communication such as e-mail, instant messaging, or social networks like Facebook and WhatsApp can provide necessary 'virtual' social support and consultation from these family members about how things are going, which could provide help with various age-related difficulties and their overall well-being. Furthermore, online sharing of A&RT platform with professional care providers and even with other patients or friends, e.g. in computer gaming, significantly increases the impact and contribution of such systems. Computer based communication gives elderly people a chance to participate in a wider part of society, making it possible to connect with people that are either similar in experience and interest or diverse, thus increasing participatory capital and possibly community commitment.

Blit-Cohen et al. [14] describes how elderly people have much to gain from this type of communication, most importantly that they can seek new information and disseminate their own ideas and share their experience with others. They can acquire new social ties and discard old ones at will. They can surf the Internet looking for relevant information and knowledge and they can find 'friends' and even support groups with people in similar conditions with whom they can share valuable experience. Being connected to the Internet and social networks does not require physical movement, which is a factor in increased social isolation among elderly people [19]. Elderly people can maintain virtual social networks from their homes so that they are not typecast according to physical appearance and that participation is not bound by time or place. The result of this is that, in contrast with the general population, the use of computer based communication by the elderly generates greater social connectivity among that group. Online participation of elderly people is also beneficiary to the society. Elderly people have a vast amount of knowledge and experience in life that they can share by using this technology and by being involved in online communities.

Not only can communication using computers stimulate social activity by the elderly. In various studies on the group-wise learning of computer A&RT use, the observation was made that being a new experience for most of them and having to go through it together stimulated social interaction within the group. Relations with healthcare providers, friends, neighbours, relatives, and work-mates that provide healthcare support as well as companionship, emotional aid, goods and services, information and a sense of belonging were observed.

In spite of the valuable potential of using social networks, a large proportion of older people do not use communications technology, the question is, why not? Some of the important barriers to older people adopting digital technologies include:

- Culture as one key factor: Health is usually defined in terms of 'disease' and older people have more diseases than younger people do. Hence, seen from a clinical perspective, the elderly suffer more illness and the solution is more healthcare. But older people themselves often view things differently. For example, frequently when asking elderly citizens about their general health, they will answer irrespective of age that their general state of health is good or very good. In other words, they enjoyed life, did not see themselves as sick and did not want to be treated [2];
- Lack of home access to the Internet: Only half of people aged 60-69 have access to the Internet at home, but this falls to 17% among the over 70s. Adults over the age of 60 are also less likely than younger adults to get Internet access in the next year. The dominant reason for not having or seeking access is that older people do not feel they need it.
- Low technology awareness: 10% of people aged 60-69 have access to the Internet but do not use it. They feel that digital technology has no relevance for them and that they would gain nothing by using it.
- Inadequate marketing: Technology marketing is generally aimed at the young, promoting gimmicky aspects of products that do not interest older people. Or, marketing is aimed at the frail elderly, a group with which older people do not identify.
- Inappropriate design: Digital equipment is designed to attract young buyers who have grown up using technology. Small buttons, fiddly controls and unnecessarily complicated interfaces can all be barriers to older, or less adept, users. The appearance of 'special' equipment is also a deterrent for some older people who do not want ugly objects cluttering up their homes.
- Anxieties: Older people tend to have certain fears regarding technology. One of them is cost. They assume, for example, that computers cost more than they actually do. Another is breaking equipment or doing something wrong. A third is security. Although many older people do not know

enough about technology to be familiar with common security problems, many know enough to be concerned.

These barriers prevent many older people from enjoying A&RT social benefits.

3.5   Homecare A&RT Media Ecology - Organizational Aspect

The current way of caring for the elderly is economically unsustainable because it is based on a costly, hospital-centered health system [2]. Other factors are structural: In most countries, primary, secondary, community and social care are organized separately with professionals who operate in an environment that encourages specialization and segregation. At best, this means that those who need care have to navigate a circuitous path through the system. At worst, it causes friction between different care providers and unnecessary expenditure on duplicate tests and services, as the elderly get shuffled from one department or organization to another. Faced with the difficulty of navigating a fragmented system, the simplest option for many is to go to the accident and emergency department.

Similarly, in most countries, funding is allocated to individual institutions rather than networks of organizations with shared goals. Each institution is a financial silo with its own income from central or local government, health insurers and patients or a mixture of the four. Many of the reimbursement mechanisms that are used also provide perverse incentives and no one agency is responsible for coordinating the healthcare that people receive or accountable for outcomes and total costs. The net effect is to direct expenditure towards the costliest part of the healthcare system: the hospital. Many older people who could be treated at home within the community and helped to live independently, end up in hospital, sometimes for quite lengthy periods of time. Yet hospitals were originally designed to isolate people with infectious diseases, not to care for those with protracted, non-communicable conditions. Cultural biases, systemic flaws and historical precedent have all driven up inefficient healthcare spending on the elderly, creating a model that is neither suitable nor sustainable.

The healthcare eco-system services for older people are provided in collaboration between hospitals, older people's care centers and primary healthcare institutes [2]. Introducing of advanced A&RT support to allow older people to live longer at home will thus have to be linked to changes in work processes and organizational practices for all of these. To achieve the benefits of new technology, extensive change ('reengineering') is often required and some-times fundamental redesign of organizations [15].

Online medical information sharing of critical health data between patients and care providers through computer based communication channels opens up a whole new arena in the healthcare eco-system (e-healthcare). This is true for routine health preventive care as well as for rehabilitation processes and acute medical response. Fast access to medical information through the computer based communication channels at the right moment is a crucial task for the whole healthcare system. In some cases the life of a patient can depend on it, but even in everyday routines, fast access to information means effective treatment and a reduction of expensive examinations and, hence, cost savings. A wearable computerized device [3] or computer rehabilitation game can pro-vide information in real time of the patient's bodily functions.

Let us assume physicians and nurses are equipped with wearable devices or rehabilitation computer games and adequate software systems. While on the move, during a ward round or during a station conference medical staff also have access to all available patient information at any time and any place in the hospital. This takes into account the mobility of medical staff; provides usable, ubiquitous access to information through connections to the clinical server and audio and visual functions. This speeds up data retrieval, improves the quality of the information and prevents patient mix-ups through context awareness. The A&RT system or device at home communicates with the infrastructure like hospital bedside displays and other devices used by colleagues, such as tablets. The technology here can improve the work of healthcare professionals: physi-cians and nurses in many aspects like improving the availability of information, presenting information in the actual context of the treatment situation, im-proving communication and knowledge sharing, and reducing the efforts for data management and documentation. Linked A&RT systems create a knowledge network. Medical information in hospitals is today available over electronic devices via the clinical information system that stores the infor-mation provided by physicians during examinations and therapy. Also, labora-tory results are delivered in large amounts to electronic medical records and can be accessed by the A&RT for healthcare workers on the move. The access to medical information is critical because the progress of treatment of a pa-tient depends on the information the physician has at hand. Fast access to stored examination results can not only result in money and time savings by avoiding repeating certain exams, but also can reduce the number of unpleas-ant and even unhealthy examinations for a patient. It is essential that all elec-tronic patient information is available even if spread over different information systems with limited communication capabilities between one another. The medical treatment process is very much distributed over the whole hospital in a sense that both patients and physicians change their locations within the hospital most of the time. As a result, equipping medical staff with wearables is promising as using a laptop often requires too much attention to be used by doctors themselves. Instead, an assistant or nurse could perform this task of computer interaction or a nurse could handle paper-based information capturing requests issued by the doctor for later entry into the hoispital information system.

4   Overcoming A&RT Media Ecology Barriers: Designing for the Elderly - User Centred Design (UCD)

As mentioned above, it is expected that A&RT applications will contribute to individual empowerment and stimulate learning and thus increase the possibility of independent living at home. Unfortunately, the vast majority of the Hi-tech industry designers are young people and sometimes it is easy for them to forget that older people exist and most of technology designers design for young people. So what will motivate the elderly, the chronically ill and the medically underserved to use interactive assistive technology systems to actively help manage their own health problems?

4.1   Concept and General Considerations of UCD

If we want to enable older people technologically, we need to help them appreciate what technology can do for them. This means tuning in to their interests, attitudes and expectations and designing programs around their needs. The belief that technology is a good thing by definition does not necessarily exist amongst older people. They need to have its value demonstrated in concrete terms with direct application to their lives. The design and development of A&RT must support the elderly end-users to overcome their resentment and enable them to accept technological aids and mobile devices without reservations. The design must then reflect the acceptance of the end users and not be the cause of new biases. That will be achieved when the elderly user are put in the heart of the process, taking into account the holistic Media Ecology environment including the three dimensions of: human, social and organizational factors.

The User Centred Design (UCD) approach ensures that during the design process of the system the human operator stays in the decision loop controlling the continuing process of action, provides feedback and makes decisions [17]. That keeps users attentive and engaged and promotes the kind of challenging practice that strengthens skills.

The focus should be on user's strengths and weaknesses and thus the software plays only a secondary role. It leverages routine functions that the operator has already mastered, issues alerts when unexpected situations arise, provides fresh information that expands the operator's perspective and counters the biases that often distort human thinking. The technology becomes the expert's partner, not the expert's replacement. A&RT software should give people room to exercise their own judgment instead of acting upon automatic, algorithmically derived suggestions. Control should shift back and forth between computer and operator running an application: by keeping the operator alert while active operations become more robust. This is sometimes not the result of automation but rather of much more effective automation that better matches the user's needs. The A&RT systems do not intend to replace human judgment with machine calculations but rather to create a decision support system that enables alternative interpretations, hypotheses, or choices. A&RT systems must use interpretive algorithms to monitor people's physical and mental states, shifting tasks and responsibilities between humans and computers. Sensing an operator struggling with a difficult procedure, the computer should initiate tasks to free the operator of distractions; capturing attention could be achieved in just the same way so that the A&RT system can actually help the user and not distract him.

For example, wearable computing uses body-worn devices providing helpful information relevant to a physical real world task. Unlike mobile computing, the focus has to be on the task and the system should not distract the user from it. To fulfill this expectation, wearable computing cannot depend on direct interaction with the wearer but needs to process contextual information from the user, tasks performed and the environment to infer the next required helpful action. This relevant information is the context of the application. That makes such systems a more powerful tool than just a mobile computing device.

The classification (taxonomy) of the overall needs of elderly people should be part of the preliminary system concept design. It has to reflect physical and emotional needs that are to be provided by social technologies and needs to support operational tasks.

Elderly care technologies basically have to satisfy three kinds of needs of the elderly; physical needs, emotional needs and task or functional needs. Physical needs include movement, locomotion, body weight assistants, robots performing cooperative tasks and tele-robotics. Emotional and mental needs are addressed by promoting sense of belongingness, encouraging the group participation, ensuring mutual care among family and friends, using games to influence elderly peoples' minds to stay positive by promoting social games as a major tool. The task needs are accomplished by facilitation using geo social applications, maps, collaboration and group meetings, messaging and interfacing the social software to the context aware systems for assisted living in order to get real time information about loved ones.

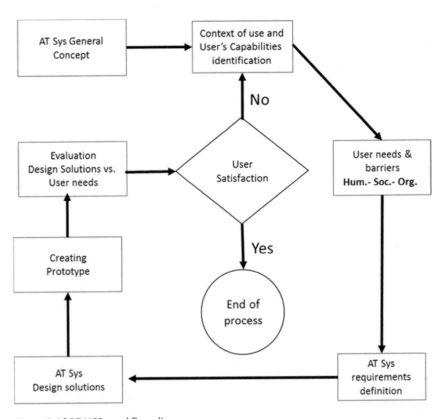

Figure 5 A&RT UCD workflow diagram

In general, UCD projects focus on getting older people online as opposed to providing them with the ongoing support they need to stay online. These projects apply technology creatively to enable older people to make connections, build social networks and actively engage in their communities for good design, including appropriate interfaces for the target group (see figure 5). In any case, UCD requires several prototyping cycles in order to best fit the system to specific users' population!

4.2 Human Factors Considerations in UCD

How old is old? It depends. Whilst avoiding using chronological boundaries, it is safe to assume that the following design consideration becomes increasingly significant after 65 years of age.

Design for older and disabled people at home has to be based on want, not an assumption of need, that much research in AT is currently aimed at exploring acceptance. Acceptability thus depends in part on the perceptions of the users themselves. Before an aid is considered, an individual has to recognize that they have a need and that it could be met.

The perceived usefulness, real needs and the feeling whether the use of the device either supports or undermines the sense of personal identity will determine the A&RT acceptance by the user. The most effective systems are those that provide routine and timely tailored clinical feedback and advice. Elderly users prefer systems that provide them with information that is specifically tailored for them and is not general in nature. Older users prefer systems that send them information on devices that fit into their normal daily routine, such as cell phones. They look for perceived potential health benefit and trust in the device, expect ease of access, use, and convenience of technology such as online peer group support bulletin boards and disease self-management tools. The anonymity and non-judgmental nature of interacting with a computer system is required.

Age related handicaps concerning vision and hearing one can reflect in design recommendations on colours, fonts, navigation, sound, content and layout and style. Examples are; that it is recommended to offer text alternatives for all non-text content [7], to use large areas of white space and small blocks of text, or to maximize the contrast between foreground and background colours [6]. When sound is used, this should be in lower frequency ranges than usual [6].

Very practical design recommendations are to avoid font sizes smaller than 16 pixels, depending of course on device, viewing distance, line height etc.

Furthermore, let people adjust text size themselves, pay particular attention to contrast ratios with text, avoid blue for important interface elements, always test your product using screen readers and provide subtitles when video or audio content is fundamental to the user experience.

Also, motor control requires special attention in design. Removing, where possible, the need to carry out complex actions using the special design enabling technologies and resolutions which help motion impaired users [11], [12]. Designs need reduced distance between interface-elements but make sure they are at least 2 mm apart. Buttons on touch interfaces should be large. Interface elements should be large to be clicked with a mouse or touched with the finger.

Learning styles of people with learning difficulties could apply to elderly users of computers, as we have seen that most elders suffer from working memory impairments. Zajicek [20] draws the conclusion that elderly people are still able to learn. Promising learning possibilities for elderly computer users are online learning courses [22] and, when learning computer skills, environments in which young and old users are mixed [21]. Playing computer games was shown to positively affect information processing, reading, comprehending and memory. It can also result in a faster reaction time and an increase in attention span and hand-eye coordination [21].This in turn can help these people in their daily lives [16]. As an additional effect, playing computer games gave nursing and support staff clues about physical and mental disabilities that had not been identified before [21]. These computer games should comply with some specific requirements, such as certain visuals and immediate feedback from the game in order to stimulate learning abilities and have to be carefully selected based on these requirements and the desired effect the game should have.

Despite this variability, three areas are particularly relevant to designing for the elderly: memory, attention and decision-making. Thus one needs to introduce product features gradually over time to prevent cognitive overload and avoid splitting tasks across multiple screens if they require memory of previous actions. During longer tasks, clear feedback on progress and reminders of goals are needed as reminders and alerts as cues for habitual actions.

One has not to be afraid of long-form text and deep content but should allow for greater time intervals in interactions, for example, server timeouts, inactivity warnings and avoid dividing users' attention between multiple tasks or parts of the screen.

Concerning decision making, one should prioritize shortcuts to previous choices ahead of new alternatives and information framed as expert opinion may be more persuasive.

Training and ongoing support helps older people overcome some of their anxieties, build skills and develop their confidence in using technology. Even if we have all the necessary hardware and software, frequently what we lack is the human element: the people and programs to deliver the necessary training and support. The training should focus on how older people want to use technology and the ongoing support should be from a trusted source.

4.3   Social Factors Considerations in UCD

The following social aspects should be taken into account within the A&RT systems UCD process:

- *Self and social identity:* The mental notion a user has about its physical, psychological and social attributes as well as its attitudes, beliefs and ideas. The self-identity is influenced by the person's membership in social groups.
- *Human interaction:* The characteristics of interaction between the wearer and another individual, as user or non-user.
- *Power and control:* The features of the socio-technical system which enables the organization or a person to control the actions of the individual wearer.
- *Trust or the trait of trusting*: Believing in the honesty and reliability of others when engaging in interaction involving exchange of information through A&RT systems and devices.
- *Gender issues:* The properties that distinguish men and women, and the implications of these properties on the design and use of wearable computing.
- *Collaboration:* The act of working jointly either between two users or between a user and a non-user.

Social computing networks and systems provide decent solutions for all the above. They can make elderly care in general and healthcare specifically a family and community wide social responsibility. The social media paradigm shift makes elderly care not only passive in technology use but also active in relationships with family members, community and care providers. Family members can take part in the elderly care while being far away from home and bridge the emotional and social distances.

Elderly care not only encompasses the healthcare but also supports their daily life such as day to day activities and operational requirements, emotional requirements, connecting to loved ones. Elderly care or assisted living systems are being developed to help and assist, monitor and prevent mishaps for the elderly, especially those who are totally dependable physically on others.

Social media is playing an important role in the life of senior citizens. Today, assistive and rehabilitation technologies are being influenced by the social network [18], which enables integration of elderly care monitoring systems for elderly, a meaningful step forwards in the connected healthcare arena. Integrated systems for enhancement of the assistive and rehabilitation living systems are already now being developed like computer multiuser games and social games designed for elderly care and rehabilitation [19], [20]. These kinds of games operate on mobile devices and focus on elderly social life. With the elderly in mind, their design is based on fewer rules and enables socializing processes with enjoyment.

Leveraging the social network capabilities with assistive technologies brings a new supporting dimension to the ageing users and this is the social inclusion of the elderly who very often suffers from isolation. Elderly people need support to keep connected, manage connections and a system that supports reciprocal care. They like to manage and/or be involved in memories and individual activities such as photo sharing [10]. Interaction is a necessity in life and physical disability is not necessarily a social disability [23]. Difference between the lifestyles of the elderly and the young is one cause of social disconnect. Internet should be active and not passive in the A&RT systems and behavioural emotional and social considerations must be taken into account as inputs for the design. [22]

Yet, a question might be raised of how social technologies get along with the already existing assistive technologies? The main issue with social networking support is whether data transmitting to social networks is trusted enough for the family members and trusted people to ensure mutual care and ensuring personal confidentiality and ethical code of conduct.

By making it an intra family social activity, the social A&RT software facilitates the old and the young to ensure mutual care. The social technology facilitates mutual responsibility and helps the family members to take along families in spite of being remote, even maintain the difference in the lifestyle and performing better elderly care. The wide spread Internet, programmable mobile platforms and apps, richer user interface and supporting tools have made it possible to develop such applications. Social networking sites are common and the transition towards internet centric applications are factors that will improve the engineering of the next generation elderly care systems .

Any elderly care assistive and rehabilitation technology integrated with social networking system should include the following social aspects in the design and development processes of the systems like the integration of all devices, especially mobile devices, into the social network, and a family centred mutual care among social network members, using the community for supporting roles. Each community member in the social network must have some advantage from being helpful to the supporting such as physician in the family, or the friend of college time for emotional support. Robotics and hardware interfacse to the social software will enable dear ones to help elderly people when they are away. Behaviour reinforcement modules and microblogging can be further reinforced by sharing funny and comic or interesting information between parents, elderly people and family members.

Elderly people are source of intellect; software must ensure privacy in intra family systems. Elderly people do not like technology as there is technology and the age gap but a user interface option must be considered to make it friendly. The level of A&RT systems integration defines the families' care support effectiveness.

4.4   Organizational Factors Considerations - Integrated Information and Shared Digital Network

Healthcare services for older people are usually provided in collaboration between hospitals, older people's care centres and primary healthcare providers. Introducing A&RT support systems allowing older people live longer at home must be linked to changes in work processes and organizational practices for all of these. To achieve the benefits of new technology, extensive change ('reengineering') is often required, and sometimes redesign of organizations [15]. The ability to deliver integrated and personalized services depends as much on integrated information systems as it does on the newly designed financing and contractual models.

Care services improve when physicians, social workers, and family have immediate, online access to information. But integrated information is not a challenge solely for providers. The elderly and their caretakers need to be proactive owners of their own health data. New technology incorporated into assistive and rehabilitation systems devices and wearables are connected to the Internet, and collects more and more data ('Big Data'), increasingly outside existing care providers networks. As such, the user becomes the central node in the use of his or her own information and therefore involved in the seamless delivery of the personalized services they need.

The digital revolution affecting our lives in almost all aspects is now disrupting the healthcare ecosystem. This wave will merge consumers and the ageing population in our case. IT technology and medical technology bring services closer to the consumer with better clinical results at much lower cost.

Unfortunately, the existing IT systems and software care providers use vary widely. Integration and cross-communication is often impossible. The reason for that is mainly suppliers' business being competition driven. It is the desire of the designers and manufacturing companies to keep their intellectual property rights (IPR). This creates numerous problems, including errors as intersystem communication disconnect. Duplicate data-entry and difficulties in comparing data from diverse sources are further problems. Much of the software that is commercially available cannot be easily adapted to reflect the requirements of individual organizations.

Four general design guidelines apply in building an A&RT systems that supports personalized service delivery:

1. The needs of each of the stakeholders must be identified: Clinical safety should obviously be a top priority, as should the security of the system itself. The information stored in electronic care records is highly sensitive, so privacy should be a key consideration for both ethical and regulatory reasons.
2. The system must work horizontally as well as vertically: Most health IT systems are designed to perform a specific set of functions in a specific department or organization. But patients move from one department to another and from one organization to another, so it is vital to build a system that spans the patient's pathway.
3. The information contained within the system must also be accurate and instantaneously available: Users must thus be able to update it wherever they are, which means that mobile access is essential.
4. The terminology and formats that different care providers and administrators use must be standardized to encourage more effective utilization of existing IT assets and minimize the amount of additional investment that is required.

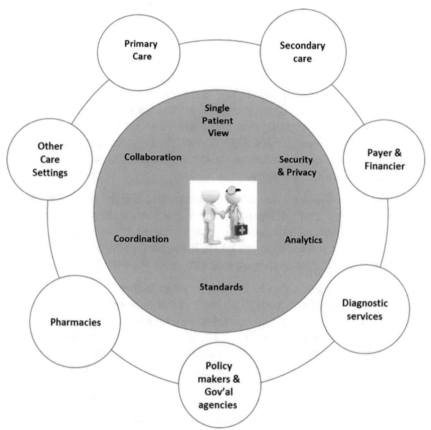

Figure 6 The connected healthcare eco-system: Personal, social, organizational aspects - Deloytte® Center for Health solutions

These guidelines will facilitate the construction of a truly interoperable client centred e-healthcare network. That, in turn, will enable multi-disciplinary service delivery teams to manage the elderly's journey more effectively, let people own and share their personal records, and pave the way for other advances based on the insights and perspectives 'Big Data' produces.

A new service delivery model for the elderly should possess several fundamental characteristics, as illustrated in figure 6. It should be far more holistic, with the emphasis on vitality and inclusion as much as on care, and focused on self-rated quality of life and wellbeing, instead of focusing on illness. It should also be organized around communities, not institutions, with clusters of care providers sharing accountability for the budgets they manage and quality of the personalized services they supply.

It should, furthermore, bring support services as close to the citizen as possible. Companies like Apple, Google and Amazon have upended retailing by taking the store to the customer and the online experiences they offer are shaping the expectations consumers carry over to other industries, as many of the companies now breaking into the healthcare business recognize. These disruptive new players are capitalizing on wireless connectivity and advanced mobile devices to erase traditional healthcare boundaries and deliver health and wellbeing services anywhere.

The new model should also reward outcomes as defined by elderly people themselves rather than activities, since it is not the number of interventions but their effectiveness that counts. For many systems this will require a shift in how we measure results. If quality of life is the goal, client experience surveys can add valuable insight on how we rate outcomes, for example.

Lastly, A&RT systems should be collaborative: Delivering individualized, integrated care entails dissolving 'the classic divide between family doctors and hospitals, between physical and mental health, between health and social care, between prevention and treatment' [11] and between private and public. Indeed, many of the factors that influence wellbeing and quality of life like nutritious food, the right housing stock, a reliable communications infrastructure and the like, lie outside the control of healthcare and social care providers. Maintaining a healthy population is not, therefore, just a job for the doctor, nurse or social worker; it is a collective challenge and opportunity for many organizations in many different industries. With the powerful disruption of new technologies and new entrants who are entering healthcare from outside industries, this collective approach empowers the elderly to co-create the necessary home health support system in a cost effective way.

The core features of a UCD healthcare business model put the individual at the heart of the system, measure and reward outcomes not activities, bring A&RT service delivery as close to the user as possible, treat health as a shared, integrated endeavour and focus on wellness and prevention, not just care and cure.

5   Conclusion

In order to effectively cope with the ageing curve of the human population worldwide, a new approach must be adopted: one in which health and wellbeing services are seamlessly connected and coordinated to meet the needs of elderly people at home, many of whom may be partially disabled, chronically ill, or have complex co-morbidities, effectively and efficiently. This calls for leveraging advanced technology enabling wireless connectivity and advanced systems and mobile devices, eliminating traditional healthcare eco system boundaries and delivering health and wellbeing services throughout the value chain.

Many older people believe they do not understand technology, are not equipped to deal with it, and do not really need to: 'technology is for the young', not for them. Yet in spite of the barriers, research makes it clear that older people are fully capable of learning to use technology and that they are interested in doing so provided they are made aware of its benefits and receive adequate training and support.

Health IT systems have the potential of empowering elderly patients to become much more active in their care process, which can not only reduce hospitalizations, emergency department use, and overall managed care costs but also deliver improve health outcomes.

To successfully incorporate interactive computer A&RT into the health care of elderly, we need to determine the impact of a wide range of barriers and drivers of health IT use, from motivation, cost, literacy, and education to language, culture, telecommunication infrastructure and access to technology. One big factor to overcome is medical personal information confidentiality, which is still an unsolved challenge in the electronic network.

Unfortunately, dealing with advanced computerized technologies, the healthcare eco system is one of the least developed and equipped environments in our lives. To leverage the huge benefits hiding in 'connected health' requires a national priority to lead a big and costly reform covering infrastructure, work protocol and procedures. This process requires leadership – political, social, business, and technological. All need to come together and collaboratively pave the way to a new era of connected healthcare Media Ecology. The need is clear, the infrastructure exists; leadership is what is needed.

6   References

[1]   IST – 2006 – 045212 SOPRANO Service oriented programmable smart environments for older Europeans
[2]   Connected and coordinated: Personalised service delivery for the elderly (Oct. 2015)
[3]   Wearable computing Michael Lawo, Robert K. Logan, and Edna Pasher (2014)
[4]   World Population Ageing, UN Population Division, (2015)
[5]   Heller, R., Jorge, J., Guedj, R. (2001): EC/NSF Workshop on Universal Accessibility of Ubiquitous Computing
[6]   Korpela et al.; Sonn et al. in Cowan and Turner-Smith, 1998 Assistive Technology Use and Stigma
[7]   Ollie Campbell Feb. 2015: Designing for the elderly – ways older people use digital technology differently
[8]   Zhao, H. (2000): Universal Usability Web Design Guidelines for the Elderly (Age 65 and Older), University of Maryland, USA
[9]   American Academy of Neurology (2003): Geriatric Neurology Fellowship Core Curriculum
[10]   Zajicek, M. (2001): Interface Design for Older Adults, Proceedings of the EC/NSF Workshop on Universal Accessibility of Ubiquitous Computing
[11]   Atkin, Jeffres, and Neuendorf in Carpenter and Buday,Computer attitudes psychology Journal of Computer Assisted Learning Vol 22(4) Aug 2006
[12]   Hirsch, T., Forlizzi, J., Hyder, E., Goetz, J., Stroback, J., Durtz, C. (2000): The ELDer Project: Social, Emotional, & Environmental Factors in the Design of Elder
[13]   Karavidas, M., Lim, N. K., Katsikas, S. L. (2005): The effects of computers on older adults, Computers in Human Behaviour, Vol. 21, pp. 697 – 711
[14]   Blit-Cohen, E., Litwin, H. (2004): Elder participation in cyberspace: A qualitative analysis of Israeli retirees, Journal of Aging Studies, Vol. 18, pp. 385 – 398.
[15]   Vimarlund V. , Olve N. Economic analyses for ICT in elderly healthcare The Academy of Management Executive 2005; 19 (4); pp. 95—108.
[16]   Oregon Health & Science University - Motivating Elderly And Chronically Ill To Use Computers To Learn How To Better Manage Health Problems (Nov. 2008)

[17] Edna Pasher et al. wearIT@work (Project # 004216) Empowering the mobile worker by wearable computing (Feb. 2005)

[18] Siewe, Y. J. (2004): Understanding the Effects of Aging on the Sensory System, Oklahoma Cooperative Extension Service, Oklahoma State University

[19] Langdon, P., Hwang, F., Keates, S., Clarkson P. J., Robinson, P. (2002) Investigating haptic assistive interfaces for motion-impaired users: Force-channels and competitive attractive-basins, Proceedings of Eurohaptics 2002 International Conference, Edinburgh

[20] Hwang, F., Langdon, P., Keates, S., Clarkson, P. J. (2001) Haptic assistance to improve computer access for motion-impaired users, Proceedings of Eurohaptics 2001, Birmingham

[21] Whitcomb, R. G. (1990): Computer Games for the Elderly, ACM/SIGCAS Conference on Computers and the Quality of Life (CQL ' 90): Proceedings, NY: ACM Press, 1990, pp. 112 – 115

[22] Browne, H. (2000). Accessibility and Usability of Information Technology by the Elderly, available at http://www.otal.umd.edu/UUGuide/

[23] Woolf, L. M. (1998): Theoretical Perspectives Relevant to Developmental Psychology, available at http://www.webster.edu/~woolflm/cognitions.html

[24] Einstein, A. and Infeld, L. (1938). The Evolution of Physics. The Growth of Ideas from the Early Concepts to Relativity and Quanta. Cambridge University Press.

[25] Owen, C. (2007). Design Thinking: Notes on its Nature and Use. Design Research Quarterly, Vol. 2, No. 1, pp. 16-27

[26] Berkun, S. (2010). The myths of innovation. 'O'Reilly Media, Inc.'.

[27] Bishop, P., Hines, A., & Collins, T. (2007). The current state of scenario development: an overview of techniques. Foresight, 9(1), pp. 5-25.

[28] Richardson, C. R., Faulkner, G., McDevitt, J., Skrinar, G. S., Hutchinson, D. S., & Piette, J. D. (2005). Integrating physical activity into mental health services for persons with serious mental illness. Psychiatric services, 56(3), pp. 324-331.

VII Intuitive Interaction
Experiences with User Groups

Marten Ellßel / Peter Knackfuß / Michael Lawo

Abstract

In this book chapter, we will reflect on our experiences gained in the context of testing solutions like those of the SafeMove[1] project. The project aimed to increase the mobility of the elderly, both near their home and on journeys. Currently, elderly people often avoid leaving their home because they feel insecure outdoors. They might have different health problems, sometimes depression and cognitive disorders. As a consequence, their reduced presence in normal daily life results in social isolation. The design of the SafeMove system is intended to encourage self-confidence in peoples' own abilities by providing home-based physical and cognitive training as well as location-based aids during outdoor life activities. The actual use of technology seemed appropriate as a potential support for this purpose. Conversely, the question arises; does the target group accept those devices and what results in an optimal user experience? Thus, it is not only about interaction but in this case about intuitive interaction. However, what makes an interaction intuitive? Interaction is intuitive when we intuitively (in advance) know how the interaction works. This is especially important when we assume people experience neurocognitive disorders at an early stage.

1	Introduction - Previous Knowledge and Interaction
2	The Expertise of Generations
3	Previous Cultural Knowledge
4	Sensorimotor Barriers of Smartphones
5	Evaluation Method
6	Test Persons
7	Experiences During the SafeMove User Evaluation
8	Conclusion
9	References

---

[1] http://www.safemove-project.eu/, accessed 9.8.2017

© Springer Fachmedien Wiesbaden GmbH, part of Springer Nature 2018
M. Lawo und P. Knackfuß (Hrsg.), *Clinical Rehabilitation Experience Utilizing Serious Games*, Advanced Studies Mobile Research Center Bremen, https://doi.org/10.1007/978-3-658-21957-4_7

1   Introduction - Previous Knowledge and Interaction

Blackler et al. explain in [1] that the intuitive use of products involves utilizing knowledge gained through other experience(s). Therefore, products that people use intuitively are those with features they have encountered before. Intuitive interaction is fast and generally non-conscious, so people may be unable to explain how they made decisions during intuitive interaction. This is in line with the statement of Mohs et al. [2] that a technical system is for a person intuitively usable as long as an effective interaction does not require any conscious previous knowledge.

As previous knowledge, we understand it involves 'left marks' of interactions within a person's life. With each interaction, we (partially) recall those left marks and apply them. Due to the duality of body and intellect, this previous knowledge always covers both. Just mental knowledge is not sufficient. The sensorimotor system of the body has to have this knowledge as automatism too.

Hurtienne [3, p33] states that, for this purpose, one needs training where the interaction to become intuitive requires repetition under as constant circumstances as possible. Any subconscious interaction depends furthermore on the spatial and temporal context.

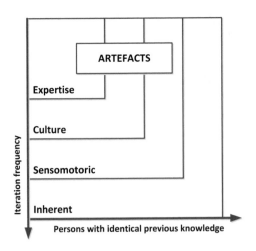

Figure 1 Continuum of previous knowledge adapted from Hurtienne [3, p34]

For Hurtienne [3, p34] previous knowledge is a continuum with four layers (see figure 1). For the classification, he uses the repetition rate of any interaction. The higher this repetition rate is, the more stable it is for the individual but also for those people having the same experiences and knowledge, the respective knowledge. As an example each person has the daily experience of vertical and horizontal and knows about it (sensorimotor layer) but only a few people know how to interact with a CAD system our how to control a power plant (expert layer).

In human-computer-interaction (HCI), the previous cultural knowledge and common experiences are an appropriate measure for the classification of a target group. During the analysis phase of the user-centred-design approach one fulfils this by interviews in the context of the application. However, this usually happens mainly on an intellectual level and thus only opens an insight into consciously applied knowledge. Statements concerning unconscious aspects people tend to state them only if they are explicitly asked (e.g. 'please explain how you tie your shoes'). In this way, aspects of the unconscious application of knowledge depend on the tutor's power of observation.

Another strategy to judge the expertise of the target group is to observe the interaction with different artefacts or mock-ups of them. These artefacts might be those of previous releases, other providers or application contexts. One expects that the subject of the experiment can apply the existing expertise to the new artefact and, in the best case, by applying unconscious previous knowledge experience within an intuitive interaction. Design guidelines like the 'Android Design' of Google or the 'iOS Human Interface Guidelines' of Apple use this approach.

When designing an interface for our target group linked to expertise there are two disadvantages: first of all the experience with an artefact in such tests has only a low repetition rate and is thus due to Nygård and Starkhammar [4] very unstable with our target group. The second disadvantage is that the previous - mainly unconscious - knowledge of our target group dates from completely different interaction concepts (e.g. switch instead of touch).

2   The Expertise of Generations

Rama et al. [5] investigated the spread of interaction styles in the Netherlands and correlated them to the formative phase of life of the birth cohort. The formative phase of adolescence and grown-ups (age 10 to 25 years old) is a key factor in the setting of the expertise on interactive systems. From their perspective, the experiences gained during this phase are so meaningful and extensive for later life that, depending on the dominant interaction style, different technology generations emerge.

Ziegler et al. [6] outline that, in 2002, people in Germany with a severe neurocognitive disorder were in the age group born between 1902 and 1937. The amount of people out of 100 with a positive diagnosis increases from 0.16 cases (age group 1938-1942 to 10.7 cases (age group before 1907). We assume that this distribution key is still valid. The people in our target group were born before 1950. This means that their formative phase of life was before 1975. According to Rama et al. [5] mechanical (until 1930) or electromechanical components dominated the interaction style in the western societies. One has to state that for people of this phase, their expertise concerning the unconscious application of previous knowledge completely differs from todays 'direct touch' interaction paradigm. Neither on the level of 'motor goals' nor on the level of 'do goals' we find sufficient commonalities.

The few existing commonalities of actual smartphones and mechanical or rather electro mechanical systems are on the level of 'motor goals'. There we have with both systems interaction conditions of switches, push buttons, control dials and throttles.

As the 'display'- interaction-style already generates increased demands on the level of 'do goals', any interaction using the 'menu style' even increases this demand. Especially, meeting the demands of planning an action does not correlate the experiences of the mechanical and electro-mechanical generation. For example, a ringing iPhone (iOS version 7) does not connect the calling party as a classical phone with dial plate does when the callee lifts the telephone receiver. Instead, the user has five options to choose: 'accept', 'ignore', 'mute', 'reminder', and 'message'; this requires a conscious planning of the interaction.

People of the mechanical and electro-mechanical generation expect a one-to-one correlation where one operation leads directly to the targeted goal. The fact that the system did not react as expected was either because the user pressed the wrong button or the system had a technical defect. By simply repeating the action, the problem was easy to fix. This familiar approach is usually not successful with systems of the 'menu style'. To repeat the action, the user has at least to understand the actual state of the system and the position within the navigation hierarchy. However, possible reasons might also be a misinterpretation of the interaction (e.g. an icon) or an incorrect planning of the action.

These aspects usually remain hidden for people of the mechanical and elec-tro-mechanical generation. In such situations, they ask questions or give statements as follows: *Which is the right button? I did everything as usual. Now it is broken.*

These statements allow an insight into the expertise concerning interactive systems. People try to transfer familiar patterns like pushing buttons in specific sequences to the 'menu' generation system. Unexpected reactions of the sys-tem are quickly interpreted as technical defects. To plan the action with an interactive system is unfamiliar for them. Thus to rethink the sequence of ac-tion is out of scope. Barnard et al. [7] confirm this strong belief in a sequential approach in their ‚walking interviews', where people of the mechanical and electro-mechanical generation use tablet computer.

To summarize: People of the generation we are working with have to really learn the new interaction concepts. For them, the underlying interaction para-digm is not intuitive. They even have to learn to overcome some of their previ-ous interaction knowledge. Both are challenging conscious cognitive tasks that require a lot of attention and repetition. This is like a barrier for the uncon-scious application of previous knowledge. In the following, we address Hurti-enne's next layer of knowledge with respect to its influence on the formative phase of life of people of the mechanical end electromechanical generation [3].

3 Previous Cultural Knowledge

In multiple ways, cultural aspects can have an influence on previous knowledge of interaction with a navigation system. On one hand, the culture influences the general ability for spatial orientation and the use of landmarks. In addition, the artefact supporting the navigation has cultural influences. In the following, we describe some remarkable aspects.

Foreman et al. [8] explain that we gain the ability for spatial orientation by moving within spaces. Thus, the structure of the landscape of the own envi-ronment influences our previous knowledge. For example, a Tyrolian in the mountains gets along better than someone from Frisia. The latter is on the other hand used to landmarks at the horizon or following the coastline. Fore-man et al. [8] further outline that people who walked a lot during their child-hood have a better orientation skill than those often going in a vehicle. Due to this, children who don't use school bus have an advantage.

Icons, signs, signboards and signposts often have historic origins (see figure 2). The design and spread differs with the culture.

Austria Netherlands Germany Hungary   Italy   Switzerland   Turkey

Figure 2 The sign for a pharmacy in different European countries [9]

In addition, the design of maps depends on cultural conventions. In the German atlas used in schools e.g. buildings are usually coloured in red, in Google maps buildings are grey. In addition, the curriculum at school and thus the way of using maps in class changes with culture and time. For example in German primary schools, the kids learned to read maps of the direct vicinity or urban hinterland. But it was only after a school reform in 1970 with an extended definition of the teaching content that global maps and thematic maps got introduced. In countries with compulsory military service, the draftees learn the basics of navigation. The economic wealth of a society further generates knowledge on navigation. People who lack the money for a journey do not need to plan travel and use a map to find their way within an unfamiliar environment like another city.

Considering these aspects for our target group of people born before 1950 and finishing their formative phase of life before 1975 we can draw the following conclusions:

Their school education was mainly before the reform of the teaching content in 1970. During their years at the primary school, they mainly learned to use topographic maps of the direct vicinity. Large scales characterize these maps showing road networks, landscape forms and buildings. The use of vehicles was low during those days compared to today. Going to school was usually by foot and not by bus. Due to this, the spatial orientation ability of those people should be higher than that of later-born generations. A main portion of the formative life phase is, at least in Germany, characterized by the economic miracle after the Second World War. It was a time when people increasingly got the travel bug and visited different places using guidebooks to explore the vicinity. This we could confirm in a user study done 2013 within the context of the above mentioned research project. Most of the guidebooks of those days were the ones of editors Karl Baedeker and Grieben as Böse outlines 1964 [10]. They contained detailed tourist maps of the city centre of the metropolises of those days. In our study, we got e.g. the following statement: 'Baedecker? Are that the red ones? Yeah, the maps were not difficult. One could just read them. There was not much about.' (Mrs. M. (born 1926).

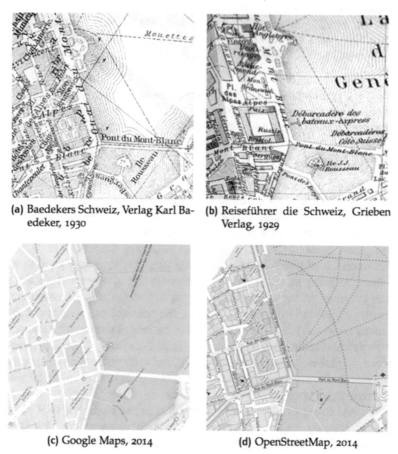

(a) Baedekers Schweiz, Verlag Karl Ba-   (b) Reiseführer die Schweiz, Grieben
edeker, 1930   Verlag, 1929

(c) Google Maps, 2014   (d) OpenStreetMap, 2014

Figure 3 Pont du Mont Blanc in Geneva on tourist maps of 1930 and today's digital maps [11], [12] taken from [13 p.82]

We will take these maps to explain the difference between actual illustrations (see figure 3). In this way, we can better understand the expectations of our target group concerning displayed maps. We take Google Maps and Open-

StreetMap for comparison with maps of 1930 published in tourist guidebooks. These maps were still in use after the Second World War until the 1960s.

The different colouring is peculiar. Maps were trichromatic, the lettering, in general, black. Areas with buildings had a red-brown colour. Water and parks were blue, streets left white. Maps of today classify the roads by colours (hiking trails, highways, motorways,...). In OpenStreetMap, points of interest have colour printing. Areas with buildings have different shades of grey in both services. In Google Maps, the classical black-and-white display format of railroads even changed.

Shades, one completely miss in maps of the 1930s; instead, one finds hatching. However, not for the road network but specific buildings in a city or highlighting points of interest like a shore area. This information one finds in Google Maps and OpenStreetMap only as text.

All buildings have the same colour. In contrast, green spaces stand out in the maps of today. The most remarkable differences concerning the lettering are the fonts, font size and tracking used. Whereas today mainly sans-serif lettering is used, the editors Baedecker and Grieben exclusively used serif lettering. The font size varies for streets, places, buildings, rivers and lakes between four and eight Didot points as Weirauch outlines [13, p37]. Compared to Google Maps [11] and OpenStreetMap [12] the fonts used in those days were much larger. Here, zooming does not change much with digital maps. Furthermore, one finds in those old maps increased tracking as soon as a street is too long for its name, whereas in the maps of today, the name of the street reprises. Digital maps often use icons with a description becoming visible when zoomed in. Old maps are often not oriented and we do not find icons there but hatching instead. Digital maps are always oriented and offer the option to use the Smartphones' build in compass. In old maps, the scale and the map section could change when turning a page; this requires the reader to perform pattern recognition and transformation. Implementations of the digital maps of today do this automatically. As the globe maps of today have an amazing range of perspectives and the time between looking at a street number and looking out from space is just a click and a few milliseconds.

To make a long story short: during the last decades, maps changed quite a lot. Of the previous knowledge of the formative age of our target group, we can transfer only few aspects. We can only use the colour of water and the marking of streets. Lettering is still often black and streets are still white. Nevertheless, for many other elements, our target group has to anticipate the previous knowledge. As long as the difference is not too large between the personal experience with maps and representation used in today's digital maps, the target group succeeds in applying the cultural previous knowledge unconsciously.

4   Sensorimotor Barriers of Smartphones

With age, the sensitivity of the musculoskeletal and perceptual human systems decreases. The interaction design affects this with respect to the motor goals. In the guidelines of Fisk et al. [14] with the title 'Designing for older adults: Principles and creative human factors approaches' one way of describing the problems being encountered is as follows: 'If it cannot be seen, heard, or manipulated, it cannot be used' ([14], p242).

The above-mentioned age related changes create increased demands for the input and output modalities of today's smartphones. An aged visual sense requires larger fonts and icons, increased contrasts, less screen reflection and brighter status LEDs. An aged hearing sense needs a higher degree of loudness and challenges the bandwidth of speakers and earphones. An aged sense of touch needs more intensive vibrations and surface sensitive switches and push buttons with distinct haptic feedback. This is a challenge when designing an interface for our target group using touch sensitive displays.

The distinct haptic feedback of the push button or a key board one replaces somehow by artificial feedback addressing the visual sense in the first instance supported by a tone and/or a vibration. This is a challenge not only for the aged human sensory receptors but also with respect to the previous unconscious knowledge concerning haptic interaction. This is even worse when taking into account that the natural skin moisture reduces with age. Barnard et al. [7] mention that since the threshold of the capacitive change needed for the interaction does not take the reduced skin moisture into account, any smooth interaction must fail. Usually application developers cannot influence this threshold. Thus, people with dry hands have to use moisture or a touch pen. Another issue is the so-called 'fat finger' problem when missing the active area of a touchscreen as the finger covers it as Wigdor et al. mention [15]. Aged hands after an intensive working life tend to be sometimes beefy with hard skin intensifying the issue.

On a touchscreen, any touch counts. One touch can have a different meaning but a repeated or a multiple one is a challenge for people with tremor. Usually the finger-icon-interaction depends on the duration and motion of the contact; a tap differs from a slide as figure 4 shows.

The very small screens of many Smartphones with slender frames and high density of active elements often clashes with the age related changed hands and fingers. The reduced sensitivity of touch and joint flexibility as tremor can cause unintended frequent touch initiating unintended interaction. In these unintended interactions where the user does not properly realize what is happening, a sufficient interpretation of the observed consequences is impossible. The use of gestures like wiping and nipping, as time dependences have to consider the changed sensorimotor abilities in order not to result in barriers due to gestures and timers.

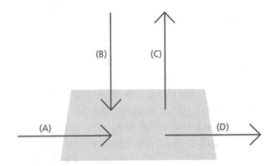

Figure 4 Touch interaction (A) slide towards (B) touch down (C) lift (D) slide down - taken from [15], p.77

When using speech interfaces one has to recognise that the style of language changes with age. Dentures or paralytic symptoms might influence the pronunciation and dialects are an issue.

5   Evaluation Method

The evaluation method seeks to empower the interaction designer for evaluating a more intuitive system. Intuition is a phenomenon mainly based on motor goals and depends on the unconscious previous knowledge of our target group. We cannot expect with our target group any expert knowledge but, as outlined above, only the cultural and sensorimotor knowledge.

To avoid the gap between the analysis and design phase, the interaction designer needs evaluation results that create a meaning for the design. The demand with respect to the evaluation method is that it provides results with subjects of our target group during the interaction with the mobile device. As intuition is an unconscious phenomenon, we have to approach the unconscious experiences of our target group. For the time being, this seems to be impossible.

However, Rohrer's Image Schemata Theory [16] provides just this. Using this language one acquires an insight into unconscious processes. It is further a design rational for artefacts and thus exactly fulfils the requirements of interaction design. Because of these dual properties, it enables a comparative assessment. Relating to the slider of figure 4, a statement like 'I tune up' means the metaphoric concept was not instantiated properly.

To close the gap between analyses and design, the described method requires completely permeating the design process. Only if one already collects image schemata and conceptual metaphors during the analysis phase can the designer pick them up to improve the design.

To collect image schemata, statements done by test persons during the interaction are sufficient. The appropriate method is the interview. In the following, we explain how to integrate our target group with this evaluation method. Here, we first look how the target group is involved in qualitative research and how to interview test persons.

5.1 Involvement of Elderly People with Mild to Moderate Cognitive Disorders

The evaluation of technical systems with our target group is rare. One exception is Prof. Dr. Louise Nygård working in the field for more than twenty years (see e.g. [4], [17], [18]). She is professor of occupational medicine at the department of neurobiology, care sciences and society of the Karolinska Institutet in Stockholm/Sweden.

In her article 'How can we get access to the experiences of people with dementia? [19], she gives the following recommendations for research in the field.

Biomedical Perspective

Firstly, she warns against taking the biomedical perspective with the diagnosis of severe neurocognitive disorders. This perspective focusses on the deficits only: a generally progressing decay, increased cognitive restrictions and the decreasing ability of finalizing activities of daily living. The biomedical perspective understands the body as place of restoration and improvements ([20], p.

21). With this approach, one separates life into phases and evaluates them with respect to the levels of risks.

This deficit-orientated perspective ignores the sensory experiences of the persons concerned and bans them from being a human. The image of people with severe cognitive disorders, due to this perspective, compares them with zombies within our society [21] and eliminates them from any research.

Socio-psychological Perspective

Yielding results is however for Nygård [19] the approach of becoming part of sensory experience of the target group. Here personal and subjective experiences of the target group need exploration. Naomi Feil is a pioneer of this approach. In the 1980s, she developed, for communication with disoriented aging people, her 'validation' called approach of which empathy is the kernel [22]. With empathy, she means gaining a deeper understanding of the experiences and life of the other person. This is distinctly different from sympathy (feel what the other feels). She describes it as walking in the other's shoes ([23], p. 15). Tom Kitwood further developed this approach for a human centred perspective on care giving. His highest goal is to keep and strengthen the personality in occupational therapy. The prerequisite here is giving people the ability to independently fulfill basic needs [24].

5.2   Statements of People with Cognitive Disorders

Nygård [19] gets access to those people via their statements. The challenge is that the interviews necessary rely on cognitive and verbal abilities. Deductive reasoning, as often required in interviews, is hard for this target group; especially with abstract topics. On one hand, this is due to difficulties when retrieving, reflecting and scheduling events and on the other hand, it is due to a changed thesaurus. Whereas the form and structure of the language remain relatively sound, the active vocabulary shrinks. Furthermore, the time for finding the right word increases significantly. Words no longer flow casually. Conversation becomes hard work as Phinney et al. outline [25].

As a consequence, statements might differ from the original intention. Especially within longer conversations due to the delays in finding the right word, oblivion and strange repetitions of topics, comments and questions. Nevertheless, these difficulties cannot be a reason to exclude people with severe cognitive disorders from research. Nygård emphasizes that there are no right or wrong answers in interviews.

A good interview gives a summary of the interviewee's perspective. It is a moot point to argue about the reliability and completeness of statements. There is always a gap between what is witnessed and what is related not only with people with cognitive disorders. The researcher contributes a great deal to bridging this gap by the interpretation of given statements. Understanding what the interviewee experiences must remain the goal of any interview. This is only successful if one accepts the perspective the interviewee has of the witnessed.

Mutual Trust

Nygård [19] recommends, inspired by ethnography, setting up mutual trust between the researcher and the interviewee as an assistive structure which supports the interviewee while explaining the personal perspective in three ways: by the context, the observation and adaptation of the interview method. The basis is mutual trust and the fact that the interviewee assents to the interviewer. The critical issues are spending a lot of time together, getting to know each other and pleasurable feelings for both.

This requires a patient researcher with sufficient time, flexibility and the will to remind the interviewees of who the researcher is, why the researcher is there and what both aim at. Only mutual trust provides a basis for obtaining seriously formal consent. The presentation of the research question as the research rational requires massive attention. The researcher has to repeat and rephrase this until the interviewee provides complete understanding as feedback.

Context

The context in which the interview takes place is also of importance according to Nygård [19]. The context may help in three ways. Depending of the context, interviewee and interviewer take different roles. The interviewer has to understand how his role might influence the interviewee. For example in the interviewee's home there could emerge the roles of host and guest. At the university, those of teacher and pupil or researcher and study subject are more probable. The actual context influences the statement on something experienced in a different (past) context. As soon as interviews take place in the natural context of the experience, they are easier to recall and more closely related to the experience. The context can further support finding the right words, as it allows the interviewee to remember. It is easier to list activities of daily life when the interviewee is in their own household. However, according to Nygård, any surplus facilitates distraction.

Observation

A well-chosen context facilitates not only the ability to remember but offers the interviewee the opportunity to also demonstrate things. Actions support the capacity for remembering in people with severe cognitive disorders. This provides an insight into issues and problem-solving strategies not consciously recognized by the interviewee and thus not possible to articulate by the interviewee. A further advantage is the spontaneous reflections of the interviewee during a demonstration. Often these reflections offer a deeper insight into the interviewee's world than just asking questions. Contextual observations provide the interviewer with a better placement of statements. What people say often differs from what they do. For the researcher, important aspects might be trivial to the interviewee and thus not mentioned, for example.

5.3   Adapting Interviews for People with Cognitive Disorders

Besides the recommendations above, the structure of an interview is of importance. Adapting this structure is, according to Nygård, important, as the style of questions, negotiation and temporal aspects of the classical interview challenge the cognitive and linguistic abilities too much by just looking at the requirements. During a verbal question, the interviewee has to realize and register the question for interpretation. Furthermore, knowledge of the domain requires the interviewee's awareness for assessment.

The answer needs a linguistic representation and, during speaking, a continuous control of accordance to the intention. Nygård recommends patience when listening and providing as much time as is needed by the interviewee for the above-described process. Time pressure increases, according to her experience, the problem of even finding the right words. The timeline of the interview has to reflect the interviewee's attention span. One can achieve this by splitting up the interview into short sections.

Any suggestion of alternate expressions in order to accelerate the interview can falsify the data. In case the interviewee loses the track, the interviewer should simply repeat the question. Here an alternative formulation of the question might be successful. One must not deviate from asking open questions as these are essential for any qualitative research. However, the content of these open questions has to reflect the typical difficulties of people with cognitive disorders. For example questions of specific temporal activities like 'what did you do today?' can be difficult. To answer this question requires recalling many events, a temporal ordering, many word retrievals of the activities and, finally, an organized verbal output. Much more appropriate are questions concerning activities usually done or opinions about specific activities of daily living. Nygård had especially good experiences with questions concerning habits and routines. These can allow access to more explicit events and valuable reflections. The interviewer's active listening and supporting feedback as enquiries is supportive for the interviewee. Such a temporary handover of the lead offers the interviewee the opportunity to evolve a story.

Furthermore, the researcher has only limited opportunities to anticipate what will happen within the native context of the interview and the event has to influence the questions. The moment for a relevant question follows the interviewee's contextualized action. Furthermore, one should not forget the interviewee's forgetfulness. Forgotten appointments, cancelled confirmations of participation or completely forgotten participations can happen. This requires the interviewer's total flexibility and talent to improvise in all situations. Only in this way, can the interviewer develop the necessary empathy to understand the world of a person with cognitive disorders.

6 Test Persons

6.1 Recruiting

The strategy to acquire test persons is, for any study in general, mission critical; the test persons determine the quality of data, the analysis and outcome. Test persons without motivation want to finalize the interview as quickly as possible, are impatient and give answers without motivation.

One approach are proxy-test persons. This is especially true for quantitative studies with its strict process in the context of people with cognitive disorders very common. However, quantitative studies are in general not appropriate due to the above-mentioned problems people with cognitive disorders have concerning abstraction and reasoning. People with an insight into the world of those with cognitive disorders answer as a proxy person the questions of the interviewing researcher. Usually the formal and informal caregivers act as proxies. In the context of human computer interaction research, specifically trained actors evaluated interactive systems (Boger et al. [26]). However, the proxy approach is inappropriate here, as the proxy has to be of the same generation and cultural background. We further doubt if the proxy can really imagine what 'being-blank', 'being-lost' or 'being-slow' situations imply for the aggrieved party.

In general, the proxy approach contradicts in two key areas the concept of embodiment:

1. Interaction is always individual and coined by the physical and social situation.
2. Creating meaning is at its core a subjective process, inseparably connected to personal needs, emotions, perception, previous knowledge and resources.

When working with a proxy, the question is, what are the criteria for inclusion or exclusion of test persons into a study? Studies from the biomedical perspective answer this question pragmatically with the medical diagnosis. In practice however, this approach is tricky. On one side, one deals as with a third person, whom a researcher cannot access without consent. Whereas a direct interview implies that the interviewee does not hide behind a mask and lives and deals with the personal diagnosis in a reflected manner. If this is not the case, one takes the risk of an ashamed or even stigmatized interviewee. However, the largest problem of the diagnosis as an inclusion criterion is its reliability. Today, there is no procedure available for diagnosis of severe cognitive disorders in an early stage (Diehl-Schmid et al., [27]). Nevertheless, general practitioners use the difficult differential diagnosis. The consequences of this became obvious in a three-year follow-up study of Pentzek et al. [28] with more than 2,000 test persons in Austria and Germany: 75% of the general practitioners' initial diagnoses were incorrect. In light of this, any self-made cognitive test must be misleading; why should a HCI researcher get better results than a general practitioner? We further doubt that with a ten minutes test like the 'Mini–Mental State Examination (MMSE)' (Kessler et al., [29]), which consists more or less of a rigid questionnaire, that the many aspects of the 'being' phenomenon could be captured. Instead of this, inclusion criteria are a consequence of the 'being' phenomenon in activities of daily mobility. This is constructive as difficulties with out of home mobility are typical and essential with cognitive disorders in the early stages. The first criterion for inclusion into the study is living in a residential care home for the elderly. This is because 77% of the informal caregivers indicate problems with out of home mobility as the reason people move to such a place (McShane et al., [30]). On the other hand, in our research we focus exactly this topic, as our application aims to delay this move. People, who primarily moved because of the 'being' phenomenon are the target group we need.

With this approach, one has to be certain that the consequences of the 'being' phenomenon were the primary reason for the move and no other reasons predominated. Another important criterion for inclusion is physical fitness. Are brawn, physical condition and balance sufficient for an out of home activity? For this purpose, one best recruits participants of the sports programs of the home as test persons. These sports programs aim to keep the ADL competences by physical training. Owing to this, one can assume that those participants are fit enough to also participate in the desired extramural activities. A good experience to access the latter test persons was the researcher's participation in such a sports program ('fit for 100' http://www.ff100.de). Institutions can access a program specifically developed for people with cognitive disorders. The training lasts for one hour, during which the participants sit in a sharing cycle. The training aims to improve ADL competences like walking, climbing stairs, keeping the balance but also the sleight of hand. Thus, the participants are a good match with our target group. The exercises aim to improve the coordination and strength of the participants. During the training the participants, the researcher gets at a glance an opportunity to judge the physical fitness. The sleight of hand ability is, for the direct touch paradigm, an important resource that the researcher could compare whilst observing the group.

We used this approach to acquire seven test persons just after the first session with the researcher's participation. However, such an approach is only possible when sufficient supported by a trusted third party, which was, in our case, the manager of the institution[2]. She had a positive attitude towards research, experiences with prior studies and understood the participation as an opportunity to benefit the development of the residents by offering a further activity.

The choice of the seven test persons used such an inclusion criterion. In an interview, the researcher identified if the exclusion criterion of having a chronic disease with intensive care as the reason of being a resident of the institution applied. This interview further checked on the motivation to participate in the study without giving details, as those might needlessly alienate the participants due to disorders related difficulties with reasoning of abstract topics. It should

---

[2] We thank Mrs. Anna Harbusch of the Bremer Heimstiftung St. Ilsabeen for her continuous support: http://www.bremer-heimstiftung.de/wohnen/stiftungsresidenz-st-ilsabeen.html

be sufficient to explain the vision and expected benefits. One needs to avoid any terms with the potential for stigmatization but concretely talk of planned activities. In our study, there were several interviews in confidence interacting with the smart phone and going for a stroll. It seemed to us important to mention that it was all at no cost, never leaving the familiar vicinity, and that the interviewee could leave the study at any time without personal consequences.

In the first interview, the motivation to participate one can easily be detected. This motivation was, in the first instance, the general social activity. They were happy about this exceptional appointment. The second motivation was to learn about smart phones. They had all seen people using such devices and wanted to know why it was so important for the individual. The motivation of two potential interviewees was to tell the interviewer about oneself. Another mainly wanted to gain personal competence by gathered information. During the first interview, it became obvious that the only topic was the identity of the interviewer and his duty within the institution. There was no interest in the study-related activities.

After a successful first interview, one can arrange a further appointment in the familiar environment of the interviewee.

6.2   Relationship Building

The mutual trust recommended by Nygård is the aim of relationship building. Only with mutual trust, can we expect an ethically acceptable consent with the interviewee. Furthermore, any familiarness helps with the 'direct touch' training and later evaluation. Both activities can have unknown, probably even scaring, aspects for the interviewee. In such situations the researcher can only moderate with mutual trust. Furthermore, regular relationship building strengthens the identity of the interviewer, making it robust for awkward 'being blank' events. In general, one can assume that the sense of shame during 'being' events is lessened with mutual trust compared to an enforced relationship.

According to Nygård, the essential steps towards mutual trust are lots of time, pleasurable feelings and a positive phase of relationship building. Nevertheless, the devil is in the time. On one side for researchers, time is an always-restricted resource. With the residents, one might expect the contrary. However, when arranging appointments one needs to take into account that daily routines and rituals are essential for the target group to give them orientation. When present, one has to respect this in any case. Even when a test person of the target group offers such a time slot, the researcher has to refuse. For a pleasurable feeling, the test person sets the meeting point. Usually it is the apartment of the interviewee in the institution. To prepare, the interviewer needs to be accustomed to the 'being' phenomenon and the role as researcher in this context. When entering the interviewee's apartment the interviewer becomes a guest and the interviewee the host. The related behaviour goes without saying.

6.3   Creating Acceptance

Accepting the ,direct touch'-paradigm means primarily reducing reservation. The time needed to overcome the existing distance between the interaction style of 'direct touch', and the previous knowledge would endanger the practicality of the evaluation method. In general, one has to assume that the interviewees of the target group have no experience with 'direct touch'. Instead of a smart phone, one should use a tablet in a first meeting to avoid spontaneous reactions like 'this is too small for me, I cannot see anything in it'. The user interface one should adapt as far as possible to the sensory motor abilities of the target group. To manage the difficulties implied by dry skin we recommend using a tablet with touch pen. With the pen, the interface can distinguish actions made by the pen from those of the hand. In this way, the user interface can ignore during pen-input touching by hand and the user can use the pen as they would during writing and rest their hand on the screen. A further benefit is that the test person can better recognize the interviewer's actions on the screen during explanations.

For the first contact, one should choose Apps whose goals on the level of 'do-goals' connect to previous knowledge of the test person and less complex on the 'motor-goals' level. For example, suitable apps are the weather forecast, classical board games, a lexicon or a picture presentation. In general, Apps without zoom or scroll- functions are beneficial as these concepts are completely new for the generation who have grown up with mechanical and electromechanical interfaces.

At the beginning, the test person must learn that interaction cannot cause any defect to the device. One should further hand it over to the test person in a switched-off mode, to experience weight and haptics. The researcher should start the introduction process with the use of an app; switching the device on, and starting the app should not be part of the demonstration. To create acceptance, one should first show how easily one could interact with the device.

Depending on the test person's reactions, one can invite the test person to interact.

In the follow-up interview, the researcher should try to get to know the test person's desires, hobbies or background for an appropriate choice of apps for a further meeting. The better the affectations of the test person one addresses, the more motivated interaction with the device becomes. As the test person more readily accepts the tablet for interaction the more the apps can approach the one destined for evaluation. One needs a stepwise approach when using apps with functions the target group is not familiar with like zooming and scrolling. The concurrent introduction of new concepts is too demanding. Switching from the tablet to the smart phone is non-critical and one can postpone this to take advantage of the larger size during the 'direct touch' training. Even a first contact on the day of evaluation can be without problems. The acceptance of tablet and smart phone is apparently replaceable. How many training sessions one needs to create sufficient familiarity with the 'direct touch' interaction is unfortunately unpredictable in our experience.

6.4   Local Context

The perspective of embodiment defines the choice of context for the evaluation. An alternative to the simple context of their own home and neighborhood does not exist. This enables, as explained above and mentioned by Nygård, the possibility to use this context as assistance for the interview. This expands the official assignment of roles for subject and researcher towards one of a guide and tourist relationship. For the researcher, this offers, at each point on a route, the opportunity to direct the conversation towards the interpretation of landmarks. The intensive sensorimotor experience of the context facilitates the test person's or rather guide's individual aided recall of the landmarks producing, more verbal material for the image schemata analysis. Beside these advantages, a familiar context reduces the feeling of uncertainty and increases engagement.

6.5   Ethical Considerations

For studies with our target group, Kim outlines that it is controversial whether persons in a later stage of cognitive disorders can give a sentential consent for their participation in research [31]. Until today, we lack binding norms and agreements. For medical and psychological research projects, one has to consult the central ethical committee of the German Medical Association or the

European Federation of Psychologists Associations. The latter is due to our experience in Austria and Switzerland having no responsibility for interactive systems. Likewise, neither the app nor its evaluation has a medical mode bias. In fact, we categorically reject any biomedical perspective.

Nevertheless, one has to respect the personal rights of the test persons sufficiently and their problems with reasoning. Nygård provides further advice on how to correctly deal ethically with the target group in research [19]. She states that it is important to come honestly, authentically and directly across, show respect and never patronize the interviewee. It is therefore sine qua non to have the mutual consent of the test person and not of a legal guardian. Only in such a way can the test persons act with respect to their own will and with the necessary respect of the researcher. As the target group has problems with reasoning it is important to describe the intended study repeatedly with different wording. The researcher has to find a way that the test person does not feel oppressed and being tested. It must be clear that the interface and not the interviewee is the subject of testing. There are no wrong answers; there is no faulty interaction. Anything has a meaning. This needs repetition until the researcher feels that the interviewee is 'fed-up' with the explanation. From then on, the researcher can assume that the test person sustainably keeps the study in mind [19].

The position paper of the Alzheimer's Society [32] also mentions the importance of a fair-mindedness. They report the great disappointment of test persons with wearable devices not knowing the purpose and functionality of these devices. Beside the possible paternalism, an ethical risk exists in building up mutual trust first, gathering research findings and later seeming to lose interest in the target group [19]. Honesty, truthfulness and reliability are key qualities in this research. Navigation systems can also have, beside the positive effects mentioned above, negative ones. A tracking functionality with an automatic notification of the caregiver can restrict the autonomy of the person looked after and even create the feeling of being supervised. According to the Alzheimer's Society [32] the advantages have definitely to be higher than the disadvantages of any interference with the human rights. Furthermore, the researcher has to take into account during development and evaluation if the privacy of the user is restricted and who can access the collected data. It is important to know if any unauthorized access to sensitive personal data of the test persons could happen.

6.6 Data Analysis

To start the data analysis, the interviews need a transcript. This is a tedious process, handicapped by the target group's form of speech. During the transcript, one notices how incomplete and inverted many sentences are. Both are hard to appreciate during the interviews. One has to extract the conceptual metaphors and image schemata out of the transcript of the interview. One needs careful study and decisions in order to identify the statements relevant for the interaction design. Relevant statements one transfers to a table. Discrete metaphors alone are unhelpful.

The context must always be in evidence. Each interviewee and interviewer receives a number for keeping privacy. In a further step, one documents and analyzed the image schemata and conceptual metaphors; Here, a translation into the subtext behind what has been said is needed. One documents similar statements only once but evaluates them every time. The documented interaction schemata and conceptual metaphors need a comparison with the interaction designer's intentions.

7 Experiences During the SafeMove User Evaluation

The first author of this chapter did the previously described research in the context of the EU funded projects SafeMove[3] and Rehab@Home[4] under the supervision of the third author at Bremen University. This took about six months in collaboration with the above-mentioned residential home care center in Bremen. The intention was to use this research for evaluation within the SafeMove project. SafeMove aims to encourage self-confidence in peoples' abilities by providing home-based physical and cognitive training as well as location-based aids during outdoor life activities.

The system developed is intended to enhance the fitness of the elderly in an interactive and pleasurable way. The system intends to support persons with cognitive disorders t in finding their way outside their home, in public traffic or at social events by helping them to remember daily life tasks like dressing themselves according to the weather conditions or to take the keys with them when leaving the house. For this purpose, two components of the SafeMove system exist: the 'atHome' module is for training and entertainment purposes

[3] www.safemove-project.eu
[4] www.rehabathome-project.eu

and the 'onTour' module for the assistance when outside the home. Further-more, the system supports caregivers by information about their clients' health condition thus keeping them healthy and mobile.

There were two test sessions, one in Linz/Austria and the other one in Kfar Saba/Israel, both via partners of the project consortium. In the following, we report the experiences gained in Israel.

7.1   The Kfar Saba Meditowers Retirement Home

Kfar Saba is a small city near the border to the West Bank. The retirement home of Meditowers there targets financially well-situated clients. There is a hospital connected but not owned by Meditowers.

Different sized apartments are available in this institution. The sheltered ac-commodations are spacious and many care providers seem to be available. The age of the clients was between 84 and 94 years. Individual Philippian nurses additionally support some of the clients around the clock.

7.2   The Test Setting

We planned to perform the tests within the apartments of the clients and par-allel tests with two clients a day for about 5 to 6 hours. We gave a detailed introduction of the system to the clients with respect to the use and different aspects of the system. This was first using the smart phone as a remote control for the 'atHome' module and a Kinect® to interact during training on routes and games. The use of features like appointments, reminders, notes, contacts, and emergency contacts was included and also those of the 'atHome' module, namely the playing of games, training of routes and use of photo galleries for describing landmarks. Using the smart phone for navigating using the 'onTour' module was a further test case.

We performed the tests in a large unused demonstration apartment. Two couples of persons with cognitive disorders and their spouse were tested for about two to three hours each day.

7.3   Performing the Tests

We started after the above-described recommended initial steps concerning e.g. the consent form with test persons and their spouses, caregivers at Medi-towers and with the interviewer team introducing the goals of the evaluation.

For each evaluation step, the interviewers first introduced the different system components. As already experienced in the above-described research during the requirement elicitation, it was not possible to keep to the schedule in the original plan.

However, all test persons tested all components except the components of appointments and reminders. Finally, six of the eight interviewees were willing to answer our questionnaires. One interviewee stated that she would not fill any questionnaire; another interviewee changed his mind beforehand and did not participate in the test.

7.4   Lessons Learned

We figured out some technical issues during the evaluation to improve the system. We had some positive experiences but also learned that some things we should avoid or respect in the further development or evaluation of the system as additional requirements.

Things to Avoid and those to Respect

Apartments in residential care homes may not be large enough to perform the games well, especially when using the Kinect. In the Meditower residential home, the interviewees with individual 24 hour caregiver support did not see any need for a system like SafeMove. Thus, these interviewees interpreted the system not as a further service but as a potential drawback with respect to their actual situation.

It is therefore important to exclude interviewees with prejudices to technology. Inflexible schedules of user tests result in frustration on the interviewees' and as well on interviewers' side; any planning requires maximum flexibility as outlined above.

The repeated oral explanations of the system by the interviewer is very important as demonstrating the use of the games e.g. by the support person. It is not sufficient to rely on the explanations displayed on the mobile device.

Positive Experiences

A very positive experience for us was that the system acceptance is independent of the age of interviewees, caregivers, relatives and medical doctors. They all accepted the navigation with landmarks and gave, in general, positive feedback for this approach. All test persons and interviewees were friendly, patient and engaged in the evaluation and testing even if they had no computer experience at all.

Technical Issues

The evaluation brought us furthermore some technical issues for improvements in a next release of the system like the fact that the display of the mobile device did not change when rotated by 180° or problems with the reminder function and scaling of the landmarks in the maps. However, all these issues were minor and we could eliminate them at short notice. In this sense, the evaluation brought us reliable results, as also known from other target groups.

8 Conclusion

Originally, we did not foresee evaluating the system with the interviewees to include the requirements elicitation process. Then, after the end of the SafeMove project, and with the improvements from the evaluation in Israel, we planned to have a further evaluation with the interviewees of the requirements elicitation. However, the time between the requirements elicitation and this foreseen evaluation was over a year. Although the commitment of the management of the residential home was still there, we could for various reasons, not involve the interviewees again.

9 References

[1]   Blackler, A. L.; Popovic, V.; Mahar, D. P.: Empirical investigations into intuitive interaction: a summary. In: MMI-Interaktiv 13 (2007), pp. 4–24

[2]   Mohs, C.; Hurtienne, J.; Kindsmüller, M.; Israel, J.; Meyer, H.: IUUI–Intuitive Use of User In-terfaces: Auf dem Weg zu einer wissenschaftlichen Basis für das Schlagwort 'Intuitivität". (2006)

[3]   Hurtienne, J.; Weber, K.; Blessing, L.: Prior experience and intuitive use: image schemas in user centred design. In: Designing inclusive futures. Springer, 2008, pp. 107–116

[4]   Nygård, L.; Starkhammar, S.: The use of everyday technology by people with dementia living alone: Mapping out the difficulties. In: Aging & Mental Health 11 (2007), Nr. 2, pp. 144–155

[5]   Rama, D. u. a.: Technology generations handling complex user interfaces. (2001)

[6]   Ziegler, U.; Doblhammer, G.: Prävalenz und Inzidenz von Demenz in
        Deutschland–Eine Studie auf Basis von Daten der gesetzlichen Kran-
        kenversicherungen von 2002. In: Gesundheitswesen 71 (2009), Nr. 5,
        pp. 281–290

[7]   Barnard, Y.; Bradley, M. D.; Hodgson, F.; Lloyd, A. D.: Learning to use
        new technologies by older adults: Perceived difficulties, experimenta-
        tion behaviour and usability. In: Computers in Human Behavior 29
        (2013), Nr. 4, pp. 1715–1724

[8]   Foreman, N.; Foreman, D.; Cummings, A.; Owens, S.: Locomotion, ac-
        tive choice, and spatial memory in children. In: The Journal of general
        psychology 117 (1990), Nr. 2, pp. 215–235

[9]   Wikipedia: Vergleich europäischer Verkehrszeichen. Version: Mai
        2014.
        http://de.wikipedia.org/w/index.php?title=Vergleich_europäischer_Ve
        rkehrszeichen&oldid=130912206, accessed May 2014

[10]   Böse, G.: Im Labyrinth der Reiseführer. In: DIE ZEIT (1964), Mai, Nr. 26,
        accessed May 2014

[11]   Google: Google Maps - Genfer See. Version: 2014.
        https://www.google.de/maps/@46.207471,6.1518305,17z, Abruf: Mai
        2014

[12]   OpenStreetMap Community: OpenStreetMap - Genfer See. Version:
        2014.
        http://www.openstreetmap.org/map=17/46.20707/6.14821&layers=C
        , accessed May 2014

[13]   Weirauch, E.: Struktur, Funktion und Entwicklung der Karten in den
        Baedeker-Reiseführern der Jahre 1827–1945. (2013)

[14]   Fisk, A. D.; Rogers, W. A.; Charness, N.; Czaja, S. J.; Sharit, J.: Designing
        for older adults: Principles and creative human factors approaches.
        CRC press, 2012

[15]   Wigdor, D.; Wixon, D.: Brave NUI world: designing natural user inter-
        faces for touch and gesture. Elsevier, 2011

[16]   Rohrer, T.: Image schemata in the brain. From Perception to Meaning:
        Image Schemas in Cognitive Linguistics, Berlin: Mouton de Gruyter,
        2005, pp. 165-196.

[17]   Nygård, L.: The meaning of everyday technology as experienced by
        people with dementia who live alone. In: Dementia 7 (2008), Nr. 4, pp.
        481–502

[18]   Nygård, L.: Living with dementia and the challenges of domestic tech-
        nology. In: Dementia, Design and Technology, IOS Press (2009), pp. 9–
        25

[19]   Nygård, L.: How can we get access to the experiences of people with
         dementia? Suggestions and reflections. In: Scandinavian Journal of Oc-
         cupational Therapy 13 (2006), Nr. 2, pp. 101–112
[20]   Innes, A.: Dementia studies: a social science perspective. Sage, 2009
[21]   Behuniak, S. M.: The living dead? The construction of people with Alz-
         heimer's disease as zombies. In: Ageing and Society 31 (2011), Nr. 1, p.
         70
[22]   Feil, N.: Validation: An empathic approach to the care of dementia. In:
         Clinical Gerontologist: The Journal of Aging and Mental Health (1989)
[23]   Feil, N., Vicki de Klerk-Rubin: Validation, ein Weg zum Verständnis
         verwirrter alter Menschen, 7. Auflage, in Reinhardts Gerontologische
         Reihe (2005)
[24]   Kitwood, T. M.; Herrmann, M.; Müller- Hergl, C.: Demenz: der person-
         zentrierte Ansatz im Umgang mit verwirrten Menschen. Huber, 2008
[25]   Phinney, A.; Chesla, C. A.: The lived body in dementia. In: Journal of
         Aging Studies 17 (2003), Nr. 3, pp. 283–299
[26]   Boger, J.; Hoey, J.; Fenton, K.; Craig, T.; Milhailidis, A.: Using actors to
         develop technologies for older adults with dementia: A pilot study. In:
         Gerontechnology 9 (2010), Nr. 4, pp. 450–463
[27]   Diehl-Schmid, P.-D. D. J.; Lautenschlager, N. T.; Kurz, A.: Gedächtnis-
         sprechstunden (Memory- Kliniken). In: Demenzen in Theorie und Pra-
         xis. Springer, 2011, pp. 419–435
[28]   Pentzek, M.; Wollny, A.; Wiese, B.; Jessen, F.; Haller, F.; Maier, W.;
         Riedel-Heller, S. G.; Angermeyer, M. C.; Bickel, H.; Mösch, E. u. a.:
         Apart from nihilism and stigma: what influences general practitioners'
         accuracy in identifying incident dementia? In: The American Journal of
         Geriatric Psychiatry 17 (2009), Nr. 11, pp. 965–975
[29]   Kessler, J.; Markowitsch, H.; Denzler, P.: Mini- Mental-Status-Test
         (MMST). In: Göttingen: Beltz Test (2000)
[30]   McShane, R.; Gedling, K.; Keene, J.; Fairburn, C.; Jacoby, R.; Hope, T.:
         Getting lost in dementia: a longitudinal study of a behavioral symptom.
         In: International Psychogeriatrics (1998)
[31]   Kim, S.: Patienteneinwilligung in der Alzheimerforschung. In: Spektrum
         der Wissenschaft Spezial Alzheimer 3 (2012), pp. 76–82
[32]   Alzheimer's Society: Position Statement - Safer Walking Technology.
         Version: July 2013.
         http://www.alzheimers.org.uk/site/scripts/documents_info.php?docu
         mentID=579

VIII Clinical Experiences from Clinical Tests Experiments in the Context of Rehab@Home

Johanna Jonsdottir / Wolfhard Klein / Rita Bertoni

Abstract

Neurological disorders, such as post stroke or multiple sclerosis (MS), are among the most common causes of long-term disability in the general population. Limitations in mobility of persons post stroke or with multiple sclerosis (MS) frequently present with limitations in reaching and grasping a n d consequently impact their independence and health related quality of life [1]. Neurorehabilitation is aimed at reducing the limitations resulting from the neurological deficit. However, not all patients have access to continuous rehabilitation. That leads to non-optimal recovery and functionality. To address this lacuna in t h e offering of necessary rehabilitation, Rehab@Home a European 7th Framework funded project was carried out. The project group came from five different countries and from several diverse disciplines, both technical and clinical, in order to implement solutions aimed at improving use of the arms of persons with neurological disorders. The project aim was to provide a motivational technological solution for rehabilitation of the arms through the creation of Serious Games with the final aim of increasing participation in life situations and quality of life of the persons. The aim of this b o o k chapter is to give an overview of how gaming has been applied to rehabilitation, the innovative solution of Rehab@Home and how the gaming platform developed has been applied to persons with neurological disorders in two different rehabilitation centres. Evaluation and treating- protocols, as well as results of the pilot efficacy study will be presented.

1	Introduction - Gaming in Rehabilitation
2	The Innovative Solution of Rehab@Home
3	Development of an Evaluation Protocol and an ICF Core Set for a Final Feasibility and Efficacy Pilot Study
4	A Pilot Efficacy Study as Validation of the Rehab@Home Gaming Platform
5	Discussion and Conclusion
6	References

© Springer Fachmedien Wiesbaden GmbH, part of Springer Nature 2018
M. Lawo und P. Knackfuß (Hrsg.), *Clinical Rehabilitation Experience Utilizing Serious Games*, Advanced Studies Mobile Research Center Bremen, https://doi.org/10.1007/978-3-658-21957-4_8

1   Introduction - Gaming in Rehabilitation

One of the goals with neurorehabilitation programs aimed at restoring functionality to a person with neurological disorder is to globally restore at the
neurological and motor level as well as at the level of interference with the
external world [17]. With more effective and innovative approaches, the hope
is that rehabilitation of these two concepts of functioning can be combined,
taking advantage of the neuroplasticity properties of the brain.

Interactive computer gaming technology has increasingly been receiving attention as a component of rehabilitation due to its motivational capacities and
its potential for offering continuity of care to persons with neurological disorders that require long term rehabilitation.

It is generally indicated that gaming technology represents a safe, feasible
and potentially effective alternative to rehabilitation one on one with a therapist, and that it may promote motor recovery after stroke [14]. In a small random controlled study on persons post stroke, an improvement in visual attention and short-term visuospatial was seen after twelve thirty minute sessions
requiring, primarily, paretic arm movements [6].

In comprehensive guidelines for stroke rehabilitation by Winstein and colleagues, virtual reality and video gaming are considered reasonable as a method for delivering upper extremity movement practice. However, the authors
emphasize the scarcity of efficacy studies and the variability of technologies
and treatment approaches which makes generalization of benefit difficult [17].

Until now Serious Games have been aimed almost uniquely at patients with
sequelae after stroke. However, the future of serious game development in
rehabilitation should include more patient groups that can use the same system setup but with adaptation of the serious gaming to the specific problems.
In this way, different patient groups can benefit from games that are motivating and repetitively train functional actions of the arm [10]. Motor symptoms
such as muscle weakness, incoordination and hypertonia affecting upper limb
function occur frequently, not only in persons after stroke, but also in persons
that have multiple sclerosis or are affected by Parkinson disorder [1], [11]. In
the Rehab@Home Serious Games approach to rehabilitation, the focus was on
developing solutions through games that could be adapted for use by patients
with different neurological disorders, a range of disability levels and at different levels of recovery.

2   The Innovative Solution of Rehab@Home

In order for the games to be effectively integrated into a clinical rehabilitation setting, it was necessary that a unified and comprehensive rehabilitation gaming platform be developed following principles of user centred design.

The Serious Games design was informed by interaction and trials with therapists and patients. Through end-user requirements of therapists and patients, the aim was to create a rehabilitation platform that would enable therapists to select and tailor games for individual patients' programs.

In the first phase of the development of the rehabilitation platform, recovered patients and health professionals at the two clinical institutes (Fondazione Don Gnocchi, Milan and NTGB, Gmunderberg) were shown potential use cases and were asked to indicate how they thought a solution could be tailored to their needs. This phase of the development was done following the Volere approach [1 2]. In this way platform and game developers were guided in developing solutions for the end user's need. The focus in this initial phase was on identifying rehabilitation paradigms for upper extremity functional deficits.

As soon as the first rough games ideas had been developed into a testable technological platform, the second phase of the project started through continuous testing of game solution on patients with feedback collected from patients and health professionals in both the clinical centres through semi structured interviews.

Through numerous single session playing of the gaming platform with patients and clinicians it was possible to arrive at solutions that were gradually nearer a serious game approach that would be able to influence patient motivation and function. Various games and various levels of difficulty were sequenced so that the session was of appropriate duration, intensity and variability to potentially lead to permanent changes in performance and motor learning.

The main requirements of health professionals and patients were followed as guidelines and the games challenged the patients through movements requiring a variety of arm and hand movements. The computer interface was developed to allow the interaction between health professionals, patients and their family, resulting in three distinct stations. One for the health professional to set up games and difficulty levels appropriate for the individual patient. One for the patient to play the games and a third one enabling patients and their families to verify progress in gaming and the related arm functions as well as to send messages to the health professionals. A calibration feature to capture the range of movements possible was introduced in order to meet the individual patient's ability.

At this point, a feasibility study was carried out to test this advanced version of the game based solutions developed in the first phase of Rehab@Home. This involved a number of recovered persons post stroke with MS or with Parkinson's Disorder in both clinical centres. The gaming was carried out in interactive sessions and lasted about one hour each. This was done to better assess if the games developed so far were sufficiently mature and ready for long-lasting use during the project's final trials.

This second testing phase focused on assessing the user experience, the motivation and the clarity of instructions provided to patients. By this means we prepared the rehabilitation sessions with the Rehab@Home patient station and games. Participants recruited in both centres were persons recovered for rehabilitation purposes.

The aim of the game environments proposed was to support movements of arms, hands over the horizontal plane (e.g., opening/closing of hand, reaching movements) including many repetitions of these tasks. The technical infrastructure deployed was kept very simple, and devices used in the feasibility phase were Kinect, Sifteo cubes and Leap Motion. They were used with a standard PC to process the input and operate the gaming environment.

Before each individual session, the patients were welcomed by the experimenter and provided with an information sheet describing the aim of the study and a consent form to be signed by the patient. The gaming environment was adapted to the individual's ability level through a calibration phase thus taking into account the actual range of motion of the arms. Whenever possible, a therapist and a family member were asked to attend the session to provide feedback on the gaming approach used. In this way feedback was available from a therapist's and caregiver's point of view.

Main results from this feasibility study indicate that the devices and games proposed to patients, caregivers and therapists were positively accepted. Several indications for improvements were provided, mainly addressing the need for calibrating the range of motion and specific needs of patients, customization of the games, and the use of visual and audio elements of the games. Motivational strategies to better engage patients, to provide feedback on progress during therapy, and to support collaborative forms of play were emphasized.

Advancements of the concept and the interface's usability for the therapists were uncovered during the prototype evaluation. Thus, by this feasibility study, it was possible to further adapt and refine the grading and motivational components of the serious gaming platform.

The final gaming platform had six games available that were played using Kinect sensors, a computer and a large screen. All the games were available in a virtual home environment where the player moved from one game to the other (see book chapter four for details).

The different games developed have different game objectives , their design and the trajectories that have to be followed to obtain its therapeutic goals. One of the games (Flowers and Bees) has two different game setups for achieving its goals. It requires the picking of flowers while avoiding getting stung by bees. In the first game setup, one needs to wait for the flowers and the bees to fall down from the top of the screen. Then one has to intercept the flowers. During the second setup, the player also collects flowers. Flowers and bees appear unpredictable on the screen. In this way, with the same game, different movements and reactions are required. Similarly the game of putting cans away in the kitchen cupboard (Kitchen game) has two ways of achieving its goal. In the first version of the game it is enough to place the hand over the can and then move it towards a similarly coloured stack of cans in one of the cupboards. In the second version of the game, the person has to place the hand over the can and then to close the hand to pick up the can. Only then can the can be moved to its proper place.

The Mad Fridge game is also played in a kitchen environment, where a crazy fridge throws out eggs and gearwheels. These objects fall down on the kitchen table and the patient needs to move his hand in horizontal and vertical direction in order to catch eggs with a basket, avoiding breaking them on the table, and to avoid gearwheels. This game tells the patient to follow indications on the screen as to where the next object would fall. Then the patient has to be ready to move the basket away if a gearwheel is thrown instead of an egg. The Blackboard mini game is also played in a kitchen environment. It shows a kitchen blackboard where shapes on the left side have to be moved to coloured spots on the right side by following a random path. Red dots on the path need to be collected. Random pairings are suggested (e.g. star - blue, square - red) on the top of the screen.

Progression of difficulty of the games is possible through requirements about greater precision of trajectories, faster velocity of movement, more goals to be reached, and so on. Scoring of the games is available through the achievement of the various objectives of the game. Objects light up when they are reached and success or failure to reach the goals have characteristic sounds. The scores are visible to the person playing and final scores are displayed prominently at the end of the game. In this way, the person playing is well informed and could compare scores from one serious game session to the other.

At the end of the session, after all games have been played, a message appears on the screen asking the player if he/she has pain or fatigue after playing, giving a choice from 0-5, where 5 was severe pain. Sometimes players indicated a minimum to moderate level of pain/fatigue (1-3). When asked whether it was pain or fatigue, they all responded that it was fatigue. This indicated that they needed to play the game with greater effort.

3   Development of an Evaluation Protocol and an ICF Core Set for a Final Feasibility and Efficacy Pilot Study

In the last decade, numerous studies have demonstrated that well planned rehabilitation can have beneficial effects on different levels of the International Classification of Functioning (ICF) [15]. Measuring the effectiveness of Serious Games interventions is crucial for the development of evidence in tele rehabilitation and virtual reality. Several measurement tools exist that can measure disease specific outcomes. However, tools that measure globally across neurological disorders are lacking [13].

Global measures that measure at all levels of the ICF are paramount for understanding the effectiveness of the approach. Further, as stated by Jette and colleagues [4] outcome measures used in tele rehabilitation should be mostly the same as those used in standard rehabilitation care. This allows the comparison across approaches. It is vital that a virtual reality serious gaming be at least as efficacious as is usual rehabilitation care.

For the final phase of validation and the efficacy study of the Rehab@Home serious gaming approach, the emphasis was put on outcome measures that are typically used and validated for persons with neurological disorders and that measure functionality and perception of health, as well as, benefits across rehabilitation approaches. Validated clinical scales of arm function and questionnaires inquiring on quality of health were used and a patient functional classification was carried out through the creation of an International Classification of Health Core Set tailored to the content of Rehab@Home. The items were selected to represent the actual health status of the patient and the core set also included all items from the general core set.

Motivation of the patients was measured through the Stroke Rehabilitation Motivation Scale (SRMS) [19] consisting of seven subscales: (1) Extrinsic Motivation Introjected, measuring motivation from external factors that result in internal pressure such as guilt or embarrassment, (2) Extrinsic Motivation Regulation, measuring motivation from external rewards such as praise by doctors or family, (3) Extrinsic Motivation Identification, measuring motivation from individual personal growth, (4) Motivation, (5) Intrinsic Motivation Knowledge, measuring motivation from learning and development, (6) Intrinsic Motivation Stimulation, measuring motivation from pleasure or personal enjoyment, and (7) Intrinsic Motivation Accomplishment.

Satisfaction with this technological approach was measured with the System Usability Scale [2] that questions the level of satisfaction with the technological approach used.

Changes in arm function were measured using the Nine Hole Peg test [8] and the Box and Block test [9]. The Nine Hole Peg test evaluates fine grasping movements while the Box and Block test evaluates grasping and passing movements. Both have been validated for persons post stroke and with multiple sclerosis [3] and [7].

Changes in perceived health status and quality of life were evaluated through the EQ-5D, including also the EQ visual analogue scale (EQ-VAS) [5] and the SF-12 [18]. The EQ-5D is a widely used, standardized, health related quality of life measure developed by the EuroQol Group to provide a simple generic assessment for use in clinical and economic studies. The EQ-5D evaluates various domains of health through five dimensions: mobility, self-care, usual activities, pain/discomfort and anxiety/depression. Furthermore, through the EQ-VAS patients were asked to show their perception of Health Related Quality of Life (HRQoL) on a scale from 0 to 100, where 0 denotes the worst imaginable HRQoL and 100 the best.

The SF-12 is a shorter version of the SF-36 giving a score of two health domains, the physical health status and the mental health status.

The final evaluation protocol and synopsis of the study were approved by the ethical committees in the two rehabilitation centres.

4   A Pilot Efficacy Study as Validation of the Rehab@Home Gaming Platform

The pilot study of the approach 'Serious Games therapeutic efficacy' was performed in 2015 and 2016 in FDGCO Italy and NTGB Austria. Included were persons post stroke and with MS with upper extremity motor deficits, able to comprehend and follow directions, not wearing pace-makers and without other co-morbidities. Inclusion criteria also included Beck's Depression Fast Screening score 4 or lower, Mini Mental State Examination Score >20, and active ROM[1] shoulder flexion >44°; elbow flexion >44°. An information leaflet was delivered to all parties that described the purpose of the study and everyone signed an informed consent form.

Participants were evaluated at the beginning and end of the study by therapists, blinded to the purpose of the study, with the aforementioned validated clinical scales for functional ability and health perceptions, with questionnaires regarding the motivational indices and satisfaction. Eventually the patients were classified using the Rehab@Home ICF core set based on the ICF.

The pilot studies were carried out differently in the two clinical centres. Therefore setup and results will be presented in the following two sub chapters.

4.1   *Intervention and Clinical Experience in Fondazione Don Gnocchi, Milan*

Fifteen persons with post stroke or with multiple sclerosis (MS) and resultant upper extremity deficits were screened for eligibility from a convenience sample at Foundation Don Gnocchi Onlus, Milan, in a period of seven months in 2015 and 2016. Ten individuals met the inclusion criteria and were recruited. The ten recruited subjects in Foundation Don Gnocchi all performed twelve individual sessions with Serious Games through the use of Kinect sensors. Each session lasted thirty to forty minutes and was supervised by a physical therapist. The game setup was done in advance of the therapy session by the physical therapists, and difficulty level and game choices were adapted to the individual patient. The game setup was revisited as needed throughout the intervention.

---

[1] ROM: Range of motion

Data Analysis

Descriptive data analysis was carried out. Data was analyzed for normality and paired t-tests and non-parametric tests were carried out as appropriate. P level was set at 0.10, and clinical improvement level was set at >10%.

Results

Ten persons post stroke (n = 2) and with MS (n = 8) completed all 12 sessions (mean age 58.73 (SD ±12.78), MS onset years 19.4 (±12.3), stroke onset years 1.2 (±0.4), 3 males/7 females). The group in general was not depressed at the beginning of treatment; scores on the Beck Depression scale ranged from 0-6 before treatment and 0-4 following treatment.

Quality of Life and Health Status

Regarding EQ100, a better perceived health status was registered in the post-trial evaluation; an increase was observed from 59.4 (21.4) % to 65.5 (23.07) %, a near clinically significant change of 10.3% although the change was not statistically significant (see figure 1). Regarding the SF12, there was an increment of five points on the mental component of the SF12, a clinically significant improvement that brought the group into a range of normality of mental wellbeing. A score of fifty or above of either domain, physical and mental, is considered a normal perception of healthy. There was no difference in the physical component following the Serious Games intervention (see figure 1).

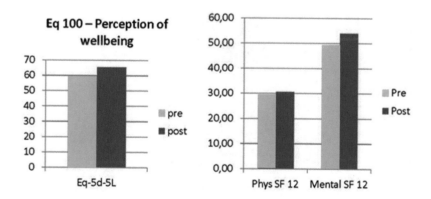

Figure 1 The EQ 100 and the SF12 scale evaluating health status and perception of wellbeing pre and post rehabilitation

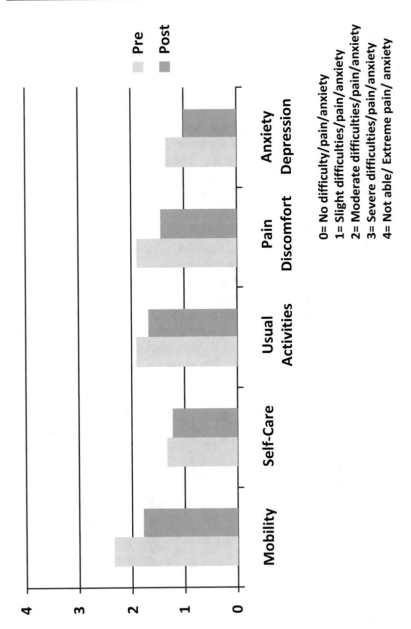

Figure 2 Mean values of patients' perception of wellbeing in the five dimensions of The EQ-5L

Results from the EQ-5D-5L (see figure 2) suggest a decrease in the level of perceived problems in all the five dimensions of wellbeing. In particular, this was evident in the dimensions regarding mobility, pain and depression, thus confirming results obtained from the Beck's Fast Screening test for depression. The group in general was not depressed before the treatment (range of scores was 0-6), neither after it (range 0-4).

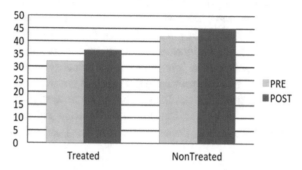

Figure 3 Mean values of patients' Box and Block test

Hand Function

There was a change in velocity of grasping and releasing movements of both treated and non-treated hands as evidenced by the Box and Block test (see figure 3). Number of cubes moved increased from 32.2 (SD 14.7) to 36.5 (15.2), a clinically significant improvement of 13.4% of the treated arm following treatment. There was a small increase of 7.8% in number of cubes moved by the non-treated hand, from 41.9 (13.3) to 45.2 (11.9), following treatment.

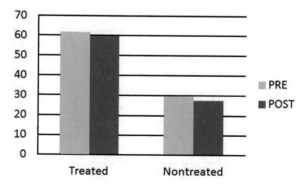

Figure 4 Mean values of patients' Nine Hole Peg test of fine hand coordination

There were no significant changes on the Nine Hole Peg test (see figure 4), a test that measures fine hand use. The participants had a small decrease in time placing the pegs from 61.3±87.3 seconds to 59.9±90.3 seconds in the treated hand, an improvement of 6.7%, and a similar trend in the non-treated hand.

In figure 5, the averaged qualifier per item (mean of all ten participants) of the Rehab@Home cores is depicted. The averaged qualifier was compared between pre to post to see if there were improvements in specific domains. In general there were small differences but they were all towards improvements in the various domains. Exercise tolerance increased enough to change qualifier status, from having a very limited exercise tolerance the participants went to having a moderate impairment of exercise tolerance (b455). This was reflected in an improvement in the activity domain of walking and moving around (d450 and d455) that before treatment was indicated as being moderately limited in the sample and after treatment was mildly limited. Also fine hand use (d440) was improved.

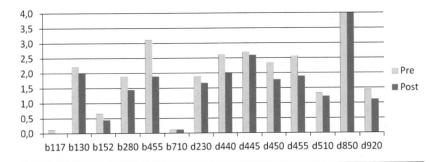

Body Functions	Activity and Participation
b117 Mental Function	d230 Carrying out daily routine
b130 Energy and drive functions	d440 Fine hand use
b152 Emotional Functions	d445 Hand and arm use
b280 Sensation of pain	d450 Walking
b455 Function of exercise tolerance	d455 Moving around
b710 Mobility of joint function	d510 Washing oneself
	d850 Remunerative employment

Figure 5 Classification with the ICF Rehab@Home Core set- Average changes in limitation – qualifiers; 0 = no limitations, 4 = total limitation

In figure 6 a typical resultant ICF profile is depicted for one of the subjects before and after the rehabilitation sessions. A reduction in impairment and difficulty is positive, a lower qualifier is better.

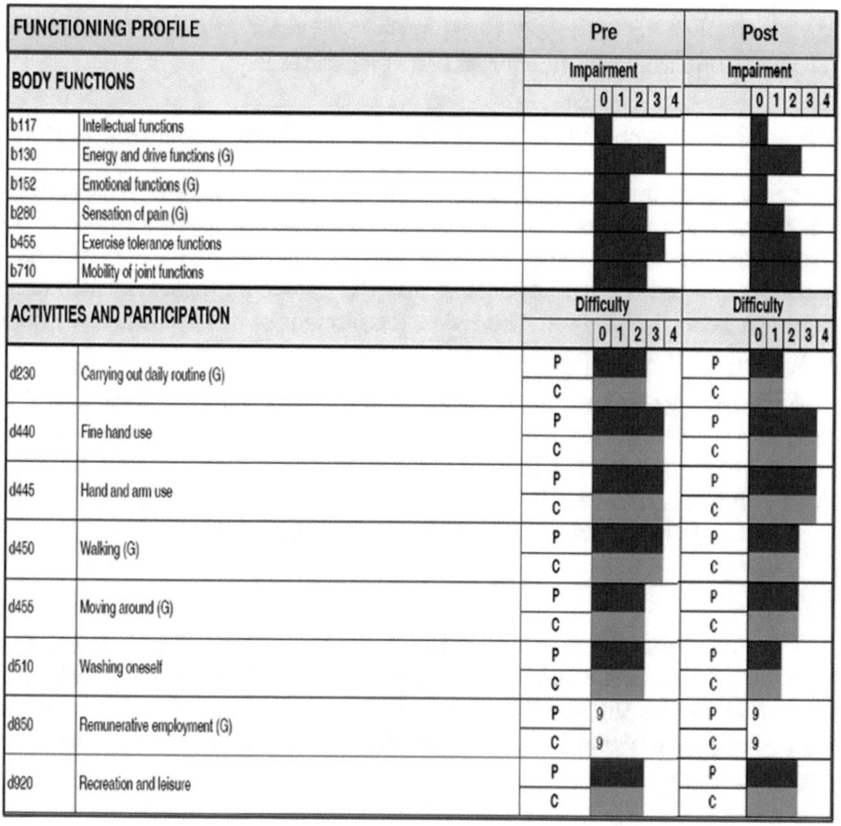

FUNCTIONING PROFILE		Pre					Post				
BODY FUNCTIONS		Impairment					Impairment				
		0	1	2	3	4	0	1	2	3	4
b117	Intellectual functions										
b130	Energy and drive functions (G)										
b152	Emotional functions (G)										
b280	Sensation of pain (G)										
b455	Exercise tolerance functions										
b710	Mobility of joint functions										
ACTIVITIES AND PARTICIPATION		Difficulty					Difficulty				
		0	1	2	3	4	0	1	2	3	4
d230	Carrying out daily routine (G)	P / C						P / C				
d440	Fine hand use	P / C						P / C				
d445	Hand and arm use	P / C						P / C				
d450	Walking (G)	P / C						P / C				
d455	Moving around (G)	P / C						P / C				
d510	Washing oneself	P / C						P / C				
d850	Remunerative employment (G)	P: 9 / C: 9						P: 9 / C: 9				
d920	Recreation and leisure	P / C						P / C				

Figure 6 A Rehab@Home ICF Core depicting the functional profile of one participant before (left) and after the Serious Games sessions (right)

In this particular participant there is a reduction in impairment of energy and motivation (b130) and in emotional functions (b152) and there is a reduction in pain (b280) and in impairment of exercise tolerance (b710). In the activity domain the patient feels less limited in carrying out daily routine (d230), in walking (d450) and in self-care such as washing herself (d510).

Motivation and Satisfaction

The user experience of persons using the Rehab@Home Serious Games system, measured after each of the twelve consecutive sessions was positive, ranging from 4 to 4.64 (1-5 Likert scale, 5 = most positive) for usability (SD±0.19), and from 3 to 4.86 for satisfaction (SD±0.51).

4.2   Intervention and Clinical Experience at NTGB in Gmunderberg/ Austria

Fifteen patients that were concordant with the inclusion criteria were recruited. Ten of them dropped out, two because they did not like the games to be played, the others because of technical problems. In the NTGB centre, the protocol set up required the patients to play the games on their own without a therapist being in the room. This led to recruitment of patients with only mild motor and cognitive impairment.

Data from five patients were included in the data analysis. Mean age was 52.8 years (± 13.88). One subject suffered from stroke which had occurred one month before the rehabilitation began. The other four subjects had the diagnosis multiple sclerosis with the mean onset time of 9.7 years (±4.6). All subjects had slight fine motor skill impairment of the right hand.

Results

The data of the Nine Hole Peg test (sec), the Box and Blocks test (count) and Grip Strength (pound) were fed to a General Linear Model MANOVA with the two within factors SIDE (left/right) and THERapy (pre/post). With Grip strength we added a third within factor (REP) for the three repetitions of measurement of grip strength.

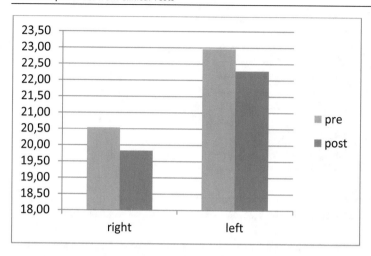

Figure 7 Mean values of patients' Nine Hole Peg test of fine hand coordination (results in Gmunderberg)

Figure 7 shows the mean values of the Nine Hole Peg test with the five patients in Gmunderberg and table 1 the output of the GLM MANOVA of Nine Hole Peg test. On the average the five patients were a little bit quicker after the therapy, but none of the effects reached statistical significance.

Table 1 Output of GLM MANOVA of Nine Hole Peg test

	SS	DgF	MS	F	p
SIDE	29,695	1	29,695	3,823	0,122226
Err	31,071	4	7,768		
THER	2,415	1	2,415	1,206	0,333853
Err	8,013	4	2,003		
S*T	0,001	1	0,001	0,002	0,968480
Err	3,270	4	0,818		

Figure 8 shows the mean values of the Box and Block test with the five patients in Gmunderberg and table 2 the output of the GLM MANOVA of Nine Hole Peg test. The patients were a little bit quicker with the right hand and with their left hand after the therapy, but none of the effects reached statistical significance.

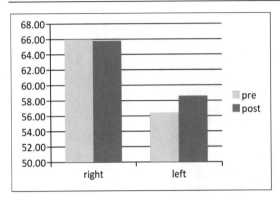

Figure 8 Box and Block test of the five patients' pre and post rehabilitation

Figure 9 shows on the left hand side the mean values of the grip strength of the five patients in Gmunderberg. On the right hand side of the same figure the values for the repetitions (wh1-wh3) are given.

Table 2 Box and Blocks test General Linear Model MANOVA

	SS	DgF	MS	F	p
SIDE	352,80	1	352,80	6,5882	0,062192
Err	214,20	4	53,55		
THER	5,00	1	5,00	1,1765	0,339077
Err	17,00	4	4,25		
S*T	7,20	1	7,20	2,4407	0,193257
Err	11,80	4	2,95		

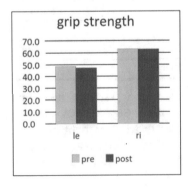

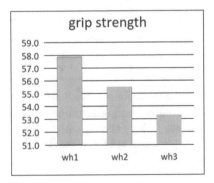

Figure 9 Grip strength of the five subjects in Gmunderberg pre and post rehabilitation

With grip strength, only the repetition factor reached statistical significance. This is a proof of plausibility in that the maximal strength can be reached just once and the second and third repetition are always lower. The right hand showed more strength but the therapy did not increase strength.

This is not very surprising since the games were not aimed at improving strength but rather the coordination of the movements. - In table 3 one finds the General Linear Model MANOVA for grip strength of the five subjects in Gmunderberg.

Table 3 General Linear Model MANOVA for grip strength of the five subjects in Gmunderberg

	SS	DgF	MS	F	p
SIDE	858,82	1	858,82	2,62515	0,180498
	1308,60	4	327,15		
THER	2,02	1	2,02	0,12210	0,744390
	66,07	4	16,52		
REP	52,93	2	26,47	8,50335	0,010475*
	24,90	8	3,11		
S*T	1,35	1	1,35	0,28322	0,622787
	19,07	4	4,77		
S*REP	3,73	2	1,87	0,87329	0,453889
	17,10	8	2,14		
T*REP	8,53	2	4,27	2,70185	0,126900
	12,63	8	1,58		
S*T*REP	0,40	2	0,20	0,12869	0,881033
	12,43	8	1,55		

*Statistically significant at P<0.05.

Clinical Experience at NTGB

In Gmunderberg no improvement as a result of the therapy could be found. This might be a problem of a ceiling effect as only slightly impaired patients participated. The games were well accepted by the participants although some patients had problems with handling the whole system. A problem encountered during the serious gaming sessions was that the Kinect sensor did not detect the movement well enough for reliable movement measurement.

This raises one of the core problems of designing computer games for neurological patients. The design of the games or the whole system must be easy to use and technically stable enough so that patients are able to use it on their own. On the other hand, the movements which have to be executed must be difficult enough to have training effects.

4.3   TMS Measurement at NTGB

Two single sessions of transcortical magnetic stimulation (TMS) were carried out on two healthy subjects during Rehab@Home game playing in order to start investigating the effect of serious gaming on cortical excitability. The method of TMS requires the placing of a short magnet field over the head of the subject. This rapidly short magnet impulse leads to a change of electrical potential within the brain and is transmitted via the corticospinal pathway to the peripheral muscles where it can be measured with EMG as a muscle response.

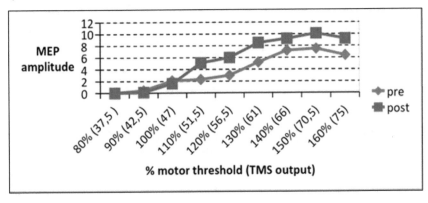

Figure 10 MEP amplitude in healthy subject 1

With both subjects, a recruitment curve was collected before and after playing one of the games for fifteen minutes. This means the strength of the magnet field is raised stepwise and this leads to a typical response curve. The recruitment curves of both subjects are shown in figures 10 and 11. After 15 minutes of playing the games cortical excitability had changed in both subjects, a smaller magnet impulse lead to a muscle response and the muscle responses were greater.

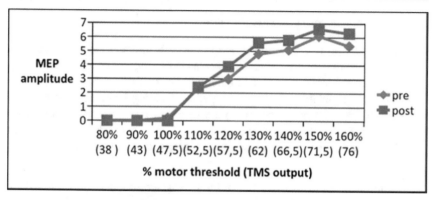

Figure 11 MEP in healthy subject 2

These initial inquiries are promising and may lead to deeper understanding of what the specific benefits of the virtual reality serious gaming approach may be. In future, measurements with TMS applied to various persons with neurological disorders during and following serious gaming rehabilitation may lead to understanding of how the neuro motor system responds to this kind of approach. Furthermore, this may allow a more precise tailoring of patients' rehabilitation.

5 Discussion and Conclusion

Fifteen people with stroke and MS finished the whole Serious Games protocol in the two clinical centres. The results of the clinical evaluation before and after the twelve rehabilitation sessions indicate small but positive changes in the functional capacity and the use of the hand following rehabilitation with Serious Games. In addition, the evaluation of the perception of health (EQ 5D-5L) and the perception of mental well-being (Short Form 12) revealed a small improvement in these areas. These results were confirmed by a reduction in the severity of the limitations in the domain of bodily functions and in the domain of activity as classified through the ICF core set Rehab@Home.

Overall, it can be concluded that the devices and the games offered to the participants in both clinical centres were well received. The integrated solution in this pilot study of Rehab@Home Serious Games was liked by patients in terms of motivation to use which coincided with a better ability of the treated arm or hand and a better perceived state of health of the patient following the intervention.

5.1 Importance of Continuity of Care

Game application to neurorehabilitation has substantially increased over the past decade with promising results, especially with regards to offering continuity of care to the various persons needing it. Up to date, continuity of care in the form of rehabilitation offered to persons exiting from hospital recovery finishing a rehabilitation cycle has been rare and costly to offer. One of the attributes of Serious Games in neurorehabilitation is its additional potential for providing motivation and goal directed exercise tasks in the home environment.

The importance of tailoring the training program to the patient's individual level of ability and constraints is paramount. Treatment plans should adapt the training load and provide distributed practice and a variability of exercises in order to improve performance and maximize the learning effect.

The ultimate goal of rehabilitation is to maximize the person's functional capacity in order to augment their chances of reintegration into social and home life.

In this project, the games were played under supervision in the rehabilitation centre, however, the majority of the persons were convinced they would continue to use it in the home setting if the opportunity arose. Amongst the multiple benefits of continuing the serious gaming rehabilitation at home under the supervision of therapists according to the patients were:

- The possibility of doing rehabilitation at home without having to spend time and resources on arriving at the rehabilitation centre at set times.
- The Serious Games approach would allow rehabilitation to be incorporated into proper daily activities with the patient free to choose the moment of the day to play.
- The availability of continuous feedback by the system and by the therapist both during and following the serious game playing was seen as encouraging for adhesion to therapy.
- The therapist supervision and constant updating of the rehabilitation program according to the real abilities of the player/patient monitored through the Rehab@Home system.

5.2 Long-term Perspective

One of the major benefits of the Serious Games approach is the possibility to continue rehabilitation under supervision of health professionals (theoretically infinitely) for much longer than would otherwise be possible. Normally patients are allowed only a set number of rehabilitation sessions covered by the health care system. With the Serious Games approach rehabilitation costs would be much lower allowing much longer therapy time. The therapy can also be incorporated into daily home and work activities thus reducing the cost to society in terms of home assistance or absence from work. Furthermore, the opportunity for longer rehabilitation time will allow bigger improvements in function. This is important for persons with neurological disorders such as multiple sclerosis and post ictus that need to consolidate the results obtained during recovery in rehabilitation centres and continue improving their function in their daily habitat.

In conclusion, it can be said that the Serious Games approach was viewed positively by all the persons playing it with no distinction of age, gender or neurological condition. The accuracy and velocity of hand and finger movements was improved following the Serious Games playing although there were some differences between centres that may be attributable to different setups of the therapy session. The results are encouraging in general, the system was fairly reliable and able to track movements of persons post ictus and with multiple sclerosis with minor to moderate arm function problems and had some efficacy in improving function and perception of health.

6 References

[1] Bertoni R, Lamers I, Chen CC, Feys P, Cattaneo D. Unilateral and bilateral upper limb dysfunction at body functions, activity and participation levels in people with multiple sclerosis. Mult Scler. 2015 Oct; 21(12): pp. 1566-1574.

[2] Bullinger HJ, F.-I.f.A.u. Organisation, and U.S.I.f.A.u. Technologiemanagement. Human Aspects in Computing: Design and use of interactive systems and work with terminals. Elsevier; 1991.

[3] Goodkin DE, Hertsgaard D, Seminary J. Upper extremity function in multiple sclerosis: improving assessment sensitivity with box-and-block and nine-hole peg tests. Arch Phys Med Rehabil. 1988 Oct; 69(10): pp. 850-854.

[4] Jette 2009: https://www.ncbi.nlm.nih.gov/pubmed/19074618

[5] Jones KH, Ford DV, Jones PA, John A, Middleton RM, et al. (2013) How People with Multiple Sclerosis Rate Their Quality of Life: An EQ-5D Survey via the UK MS Register. PLoS ONE 8(6): p. e65640.

[6]   Kim BR, Chun MH, Kim LS, Park JY. Effect of virtual reality on cognition in stroke patients. Ann Rehabil Med. 2011 Aug; 35(4): pp. 450-459.

[7]   Lin KC, Chuang LL, Wu CY, Hsieh YW, Chang WY. Responsiveness and validity of three dexterous function measures in stroke rehabilitation. J Rehabil Res Dev. 2010; 47(6): pp. 563-571.

[8]   Mathiowetz V, Kashman N, et al. Adult norms for the Nine Hole Peg Test Of Finger Dexterity. OTJR: Occupation, Participation and Health.1985; 5(1): pp. 24-38

[9]   Mathiowetz V, Volland G, Kashman N, Weber K. Adult norms for the Box and Block Test of manual dexterity. Am J Occup Ther. 1985 Jun; 39(6): pp. 386-391.

[10]   Proffitt RM, Alankus G, Kelleher CL, Engsberg JR. Use of computer games as an intervention for stroke. Top Stroke Rehabil. 2011 Jul-Aug; 18(4): pp. 417-427.

[11]   Quinn L, Busse M, Dal Bello-Haas V. Management of upper extremity dysfunction in people with Parkinson disease and Huntington disease: facilitating outcomes across the disease lifespan. J Hand Ther. 2013 Apr-Jun; 26(2):c148-54; quiz 155.

[12]   Robertson S and Robertson J. Mastering the Requirement process: getting requirements right. 3rd Edition. Pearson Education, Inc. 2013

[13]   Salter KL, Teasell RW, Foley NC, Jutai JW. Outcome assessment in ran-domized controlled trials of stroke rehabilitation. Am J Phys Med Re-habil. 2007 Dec; 86(12): pp. 1007-1012. Review.

[14]   Saposnik G, Levin M; Outcome Research Canada (SORCan) Working Group. Virtual reality in stroke rehabilitation: a meta-analysis and im-plications for clinicians. Stroke. 2011 May;42(5):1380-Teasell R, Foley N, Salter K, Bhogal S, Jeffrey Jutai J, Speechley M. EVIDENCE-BASED REVIEW OF STROKE REHABILITATION. Executive Summary (14th Edi-tion) 2012. www.ebrsr.com

[15]   Teasell R, Hussein N, McClure A, Meyer M. Stroke: More than a 'brain attack'. Int J Stroke. 2014 Feb; 9(2): pp. 188-190.

[16]   Winstein CJ, Stein J, Arena R, Bates B, Cherney LR, Cramer SC, Deruyter F, Eng JJ, Fisher B, Harvey RL, Lang CE, MacKay-Lyons M, Ottenbacher KJ, Pugh S, Reeves MJ, Richards LG, Stiers W, Zorowitz RD; American Heart Association Stroke Council, Council on Cardiovascular and Stroke Nursing, Council on Clinical Cardiology, and Council on Quality of Care and Outcomes Research. Guidelines for Adult Stroke Rehabilitation and Recovery: A Guideline for Healthcare Professionals From the American Heart Association/American Stroke Association. Stroke. 2016 Jun; 47(6): pp. e98-e169.

[17]   Ware JE, Kosinski M, Keller SD. A 12-Item Short-Form Health Survey: construction of scales and preliminary tests of reliability and validity. Med Care 34: pp. 220- 233, 1996.

[18]   White GN, Cordato DJ, O'Rourke F, Mendis RL, Ghia D, Chan DKJ. Validation of the Stroke Rehabilitation Motivation Scale: a pilot study. Asian J Gerontol Geriatr 2012; 7: pp. 80–87.

IX Recognizing Emotional States
An Approach Using Physiological Devices

Ali Mehmood Khan / Michael Lawo

Abstract

Recognizing emotional states is becoming a major part of user context for wearable computing applications. The approach presented here starts from the research hypothesis that a wearable system can acquire a user's emotional state by using physiological sensors. The purpose is to develop a personal emotional states recognition system that is practical, reliable, and can be used for health-care related applications. We use, as book chapter three described, the eHealth platform [1] which is a ready-made, light weight, small and easy to use device. The intension is to recognize emotional states like 'Sad', 'Dislike', 'Joy', 'Stress', 'Normal', 'No-Idea', 'Positive' and 'Negative' using a decision tree classifier. In this chapter, we present an approach that exhibits this property and provides evidence based on data for eight different emotional states collected from 24 different subjects. Our results indicate that the system has an accuracy rate of approximately 98%.

1	Introduction
2	Related Work
3	Experimental Set-up
4	Results and Analysis
5	Conclusions and Future Work
6	References

© Springer Fachmedien Wiesbaden GmbH, part of Springer Nature 2018
M. Lawo und P. Knackfuß (Hrsg.), *Clinical Rehabilitation Experience
Utilizing Serious Games*, Advanced Studies Mobile Research Center
Bremen, https://doi.org/10.1007/978-3-658-21957-4_9

1 Introduction

It is hard to express your own emotions; no one can accurately measure the degree of her or his emotional state in a specific situation. However, according to Darwin, '....the young and the old of widely different races, both with man and animals, express the same state of mind by the same movement" [16]. According to Paul Ekman, there are seven basic emotions which are fear, surprise, sad, dislike, disgrace, disgust and joy [14]. The concept behind emotional states (also known as *affective computing*) was first introduced by Rosalind Picard in 1995 [2].

Since then, the affective computing group has produced novel and innovative projects in that domain [3]. Emotional states recognition has received attention in recent years and is able to support the health care industry. Emotions and physical health have a strong link in influencing the immune system, too [15]. Occurrence of an emotional disorder is in more than 50% of the cases untreated, chronic stress [6]. According to Richmond Hypnosis Centre, 110 million people die every year as a result of stress. That means that, every 2 seconds, 7 people die [4]. According to American Psychological Association, in 2011 about 53 percent of Americans claimed stress as a reason behind personal health problems [5]. According to WebMD (www.webmd.com), intense and long term anger causes mental health problems including anxiety, depression, self-harm, high blood pressure, coronary heart disease, colds and flu, stroke, gastro-intestinal problems, and cancer [6]. The Occupational Safety and Health Administration (OSHA) reported that stress is a threat for the workplace. Stress costs American industry more than $300 billion annually [6]. Stress is a huge problem for mankind [9] as stress affects our health negatively, causing headaches, stomach problems, sleep problems, and migraines [13]. It can also cause many mouth problems, the painful TMJ (temporomandibular joint) syndrome, and tooth loss [7]. Just citing some statements concerning the influence of emotions illustrates the overwhelming importance of monitoring the stress level of a person: 'Stress has an immediate effect on your body. In the short term, that's not necessarily a bad thing, but chronic stress puts your health at risk" [8]. 'If you have a destructive reaction to anger, you *are more likely to have heart attacks*" [12] whereas '*an upward-spiral dynamic continually reinforces the tie between positive emotions and physical health*' [17].

Modern day lifestyles have led to various physical and mental diseases such as diabetes, depression and heart diseases as well. Although the negative effects of stress are known to people, they choose (deliberately or otherwise) to ignore it. They need to be forcefully notified that they must shrug off negative emotions; this can be by either sending them calls or some video clips, text messages or engaging them in games [10]. Emotions are the feelings which influence the human organs. There are medical studies showing that negative thinking or depression can adversely affect your health [19]. Probably automatic and personal applications can be very helpful if they can monitor one's emotional states and persuade people to come out of negative emotional states. According to William Atkinson *'The best way to overcome undesirable or negative thoughts and feelings is to cultivate the positive ones"* [18].

Emotional recognition technology can tackle this problem as it is able to monitor an individual's emotional states. This kind of system can also send an alarm call to a person when in a negative emotional state for a long time or notify the professional caregiver or family member. Such a system could log an individual's emotional states for later analysis. In some cases, especially for people with heart diseases, emotional states are required along with physical activities performed. Physiological information doctors also need to know for diagnostic purposes the patient's conditions outside the doctor's clinic [11].

Emotional computing is a field of human computer interaction where a system has the ability to recognize emotions and react accordingly. We thus wanted to recognize emotional states using physiological sensors which should be able to identify a few of them like sad, dislike, joy, stress, normal, no-idea, positive and negative. Our research aims to prove that one can detect the aforementioned emotional states by only using the data collected by physiological sensors.

2   Related Work

There is quite some research available on using automated systems for recognizing emotional states: There are systems using speech ([23], [24], and [25]), facial expressions ([26], [27], and [28]) and physiological devices ([20], [21], [22], [29], and [30]). In our research on wearable systems, we solely used body worn physiological devices measuring e.g. electromyography (EMG), blood volume pulse (BVP), galvanic skin response (GSR) and skin temperature.

Another intended output of research is to recognize specific emotional states like being sad ([20], [21], [22] and [30]), or happy and feeling joy ([20], [21], [22], [30] and [31]), or in a normal or neutral emotional state ([21], [30] and [31]), or in a negative one [29]. Different physiological devices were used for this purpose like Electroencephalography (EEG), galvanic skin response and pulse sensor to detect joy, anger, sad, fear and relief. Audio and visual clips were used as a stimulus for eliciting the emotions [20]. With electrocardiography (ECG) happiness, sadness, fear, surprise, disgust, and a neutral mood were recognized using audio and visual clips as a stimulus for eliciting the emotions [21].

To recognize joy, anger, sadness and pleasure in a study, music songs were the stimulus for eliciting the emotions and electrocardiography, electromyography, skin conductance, and a respiration sensor the detector [22].

With data from the blood volume pulse, electromyogram, respiration and skin conductance sensor twenty experiments in twenty consecutive days, testing around 25 minutes per day on each individual, a study figured out neutral, anger, hate, grief, love, romantic, joy and reverence as emotion states. The classification accuracy among the eight states was 81% [31].

From those studies we got to know that different techniques can be used as a stimulus for eliciting the emotions: pictures, video clips, audio clips, games etc. An established methodology widely used in experiments studying emotion and attention is using the International Affective Picture System (IAPS) for stimulation (http://csea.phhp.ufl.edu/media.html). To maintain novelty and efficacy of the stimulus set, the IAPS images themselves are typically not shown in any media outlet or publication and thus also not shown here[1],[2]. The International Affective Picture System (IAPS) provides normative emotional stimuli for emotion and attention under experimental investigations. There is a large set of emotionally-evocative, standardized, colour photographs, internationally accessible and available that includes contents under semantic categories. The IAPS (pronounced eye-aps) is being produced and distributed by the Center for Emotion and Attention (CSEA) at the University of Florida [32] and used in many studies (e.g. [33], [34], [35], [36], [37], [38], [39]).

In our work, we used, as stated above, solely the four physiological sensors of blood volume pulse (BVP), galvanic skin response (GSR), electromyography (EMG) and skin temperature in order to recognize the emotional states of

---

[1]   The IAPS may be received and used upon request by members of recognized, degree-granting, academic, not-for-profit research or educational institutions (http://csea.phhp.ufl.edu/media/iapsmessage.html).

[2]   An example can be found on Research Gate in Courtney, C.G., et al., Better than the real thing: Eliciting fear with moving and static computer-generated stimuli, Int. J. Psychophysiol. (2010), doi:10.1016/jijpsycho.2010.06. 028

stress, joy or happiness, sadness, normal or neutral, and dislike. As the aim was to use as few sensors as possible for a reliable output, we also evaluated different combinations and quantity of sensors. Furthermore, we placed, due to user acceptance considerations, the sensors only on the left arm in contrast to the common approach where sensors are placed on different parts of the body. With the IAPS we intend to stimulate the emotions for the ground truth for our research hypothesis that by using only data from the above mentioned four physiological sensors one can reliably detect an emotional state.

3 Experimental Set-up

The eHealth Sensor platform [1] described in chapter three of this book connects with the Raspberry Pi [41] the four physiological sensors placed on the lower left arm of the subject sitting in front of the flat screen of a computer. The stimulating application presents the IAPS to change the emotional state and the sensors collect the sensor data related to the stimulus. Figure 1 shows the placement of the sensors (a), the eHealth Sensor platform (b) the Raspberry Pi (c) and the pulse sensor (d).

The e-Health Sensor Platform has been designed by Cooking Hacks (the open hardware division of Libelium) in order to help researchers, developers and artists to measure biometric sensor data for experimentation, fun and test purposes. Cooking Hacks provides a cheap and open source alternative compared to the proprietary and price-prohibitive medical market solutions. However, as the platform does not have medical certification, it cannot be used to monitor critical patients who need accurate medical monitoring or those whose conditions must be accurately measured for an ulterior professional diagnosis. [1]

Our research exclusively intendeds to show the feasibility of the approach but with no medical study of any kind.

IAPS images [32] are clustered in different categories as 'Pleasant', 'Neutral', 'Unpleasant', 'Mutilations', 'Attack', 'Household Objects', 'Families', 'Erotica', 'Non-threatening animals', 'Neutral people', 'Neutral scenes' and 'Snakes'. To be compatible with the state of the art in our domain we defined the following groups: 'Sad', 'Dislike', 'Joy', and 'Stress'.

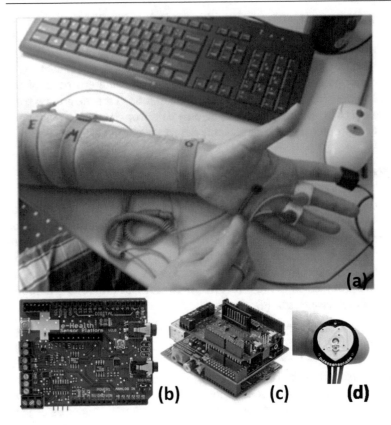

Figure 1 Placement of sensors (a), eHealth Sensor platform (b), Raspberry Pi (c), Pulse sensor (d)

Some merging of categories was required: 'Pleasant' we merged with 'Joy' and 'Neutral' and 'Household Objects' we put in a new category 'Normal' as none of the emotions we were interested in. 'Unpleasant' and 'Negative' we put to 'Sad' and 'Dislike'. 'Mutilations', 'Attack' and 'Snakes' we merged to 'Stress'. In this way we tried to harmonize the categories with those known from the literature as described above in the Related Work chapter.

Our application, with a predefined start and end, shows IAPS images in a sequence in order to change observer's emotional states within an experiment. The intention was to capture a ground truth for the experiment by bringing the IAPS indication of emotion with the observer's indication in line and assigning the different sensor data simultaneously to the emotional state for later judgment of the emotional state solely from the measured sensor data.

We chose 100 IAPS images from different categories and presented them in five iterations. The images were shown as a slide show with five different images from a specific group and a timer of 5 seconds for each image. A Latin square sequence was used for the experiments as shown in table 1.

Table 1 IAPS image sequence for experimental setup

Sad	Dislike	Joy	Stress
Dislike	Joy	Stress	Sad
Joy	Stress	Sad	Dislike
Stress	Sad	Dislike	Joy

In addition to the personal information of age, gender, height and weight, observers had to choose one emotional state out of the six categories of 'Sad', Dislike', 'Joy', 'Stress', 'Normal', or 'No Idea' after each five images of a group. Although only images of the four groups of 'Sad', Dislike', 'Joy', and 'Stress' were shown, the observer could also choose the last two categories. The application generates a text file with the observer's feedback and the IAPS images with a timestamp.

The IAPS image numbers for the training setup were the following [32]:

'Dislike' (35) 3210, 3500, 3530, 3550, 5120, 6190, 6200, 6210, 6211, 6212, 6230, 6250, 6260, 6300, 6350, 6370, 6510, 6821, 6830, 6831, 6834, 6838, 9250, 9250, 9254;

'Joy' (26) 1440, 1500, 1590, 1600, 1850, 2250, 2304, 2510, 2560, 5300, 5890, 7200, 7260, 7280, 7284, 7352, 7410, 7460, 7481, 8090, 8116, 8280, 8320, 8400, 8465, 8510;

'Sad' (25) 3210, 3500, 3530, 3550, 5120, 6190, 6200, 6210, 6211, 6212, 6230, 6250, 6260, 6300, 6370, 6510, 6821, 6830, 6831, 6834, 6838, 9070, 9250, 9252, 9254;

'Stress' (25) 1019, 1052, 1070, 1080, 1090, 1110, 1111, 1113, 1300, 1930, 3030, 3071, 3080, 3101, 3110, 6370, 6510, 6821, 6830, 6831, 6834, 6838, 9250, 9252, 9254.

In the test setup we used the following images:

'Dislike' (25) 1111, 1201, 1280, 1300, 1303, 1930, 2691, 2730, 2750, 9006, 9008, 9040, 9090, 9290, 9300, 9341 ,9432, 9440, 9470, 9480, 9490, 9592, 9611, 9912, 9921;

'Joy' (25) 1340, 1463, 1750, 1920, 2070, 2165, 2208, 2340, 2341, 2360, 2791, 4611, 4641, 4651, 4652, 4653, 5600, 5621, 7330, 8080, 8120, 8180, 8200, 8370, 8420;

'Sad' (25) 2120, 2205, 2520, 2590, 2800, 3180, 3181, 5970, 5971, 6020, 7380, 9001, 9010, 9040, 9041, 9102, 9180, 9470, 9560, 9561, 9611, 9622, 9800, 9912, 9921;

'Stress' (25) 1050, 1051, 1114, 1120, 3000, 3010, 3015, 3051, 3053, 3060, 3064, 3068, 3069, 3100, 3102, 3120, 3130, 3150, 3168, 3170, 3266, 3400, 6540, 9181, 9405.

The experiments were conducted in a calm room without noise or destraction and at normal temperatures at the Automation Engineering School of the University of Electronic Science and Technology of China in Chengdu during July and August 2014. To make sure that the readings from the galvanic skin response sensor (GSR) were accurate we asked the participants to dry their hands with a dryer before beginning the experiment. Since GSR measures sweat glands as well, moist hands would result in an erroneous result. To ensure full concentration from the participants, the light in the room was kept very low and we also asked them to turn off their mobile phones during experiments. Participants were asked to wear sensors on their left arms, palms and fingers (as shown in figure 1 (a)). We recruited 26 participants (21 males, 5 females) for our experiment; two of them could not complete the experiments so we ended up with 24 participants (19 males, 5 females). The range of participants' age was from 20 to 44 (mean 26.17, SD 5.14) and the body mass index (BMI) of the participants was between 18.7 and 26.6 (mean 21.44, SD 2.17).

The participants had to perform the experiments twice; the first experiment was useful in getting the participants to familiarize themselves with the setup. The participants should get used to the setup, as from pre-test we knew that participants tended to be a little bit nervous due to the unknown physiological devices and the long cables; we wanted to avoid a bias from this influencing our data. No result from the first experiment was used for analysis. Only the second attempt was used for analysing their data and the two experiments took place on different days.

Thus, in the second experiment all participants already knew about the set-up and they were not hesitating with the sensors. They performed the task with confidence and their data was stored for later analysis. We used the same settings for both experiments but the IAPS images were different; we showed participants as mentioned above the different IAPS images changing their emotions to sad, dislike, joy and stress. After showing a set of five images from one group participants had to choose from the questionnaire one of the six (four plus two) emotional states. The physiological data from the connected sensors was logged on a laptop with a time stamp together with the image numbers and respective category. Both files merged generated the file for post analysis.[3]

The experiments with the 24 participants had a duration of eleven to twelve minutes each. The sampling rate was around 650 Hz. A window of five seconds was used to normalize the data.

4   Results and Analysis

The first attempt was to identify the ground truth. For this purpose we looked for the reliability with which participants were able to assign the individual sensation to the IAPS image category. In table 2 the results of this assignment are summarized.

Table 2 Assigned emotional states due to IAPS image category

Emotional states	Frequency	Comments
Sad	21/24	'Sad' was ignored by 3 participants
Dislike	24/24	'Dislike' was chosen by all participants
Joy	24/24	'Joy' was chosen by all participants
Stress	20/24	'Stress was ignored by 10 participants
Normal	14/24	'Normal was ignored by 10 participants
No idea	10/24	'No-Idea was ignored by 14 participants

Only four of the participants chose all the IAPS images given emotional states correctly. This was due to the fact that it was hard for the participants to distinguish between sad, dislike and stress. Also being asked to distinguish

---

[3] The data can be accessed via the link: https://github.com/alikhan1982/Affective-Computing/tree/784ce61641ca39133c6a5f5f9ebd213a47e0b6ca/8

between joy and normal during experiments was not a straightforward task. That also explains why some emotional states were ignored by participants (as shown in table 2).

'As everyone knows, emotions seem to be interrelated in various but systematic ways: Excitement and depression seem to be opposites; excitement and surprise seem to be more similar to one another; and excitement and joy seem to be highly similar, often indistinguishable" [43].

Based on this finding we generated another dataset (Type 2) from our experimental data by using only two emotional categories: **'Positive'** by merging 'Joy' and 'Normal' and **'Negative'** by merging 'Sad', 'Dislike', and 'Stress'. Furthermore, we excluded 'No-Idea'. This merge was also done with the acquired dataset. The first dataset (Type 1) contains 'Normal', 'Sad', 'Dislike', 'Joy', 'Stress' and 'No-Idea', whereas the second dataset (Type 2) contains only 'Positive' and 'Negative'.

For the analysis we used WEKA *(www.cs.waikato.ac.nz/ml/weka/)*, a collection of machine learning algorithms for data mining tasks. Due to the fact that it was a huge dataset, it was not possible to process the data of all 24 participants at a time. Therefore, we divided our dataset into six groups, each group containing the data from four participants (as shown in table 3); we grouped the four participants who correctly chose all emotional states in the first group; others were assigned in alphabetic order to the remaining groups.

Table 3 Groups of participants for data mining purposes

No.	Age	Gender	BMI	Chose Emotional states (Type 1)
1	25,24, 25,26	3 Males, 1 Female	23.4, 20.5, 20.8, 21	Normal (4), Sad (4), Dislike (4), Joy (4), Stress (4) and No-Idea (4)
2	24,25, 25,38	4 Males	23, 23.9, 21.2, 26	Normal (0), Sad (3), Dislike (4), Joy (4), Stress (4) and No-Idea (2)
3	24,24, 25,44	3 Males, 1 Female	19.1, 20.8, 26.6, 19.4	Normal (3), Sad (3), Dislike (4), Joy (4), Stress (4) and No-Idea (1)
4	20,25, 25,33	2 Males, 2 Females	20.5, 20.2, 18.7, 20	Normal (2), Sad (4), Dislike (4), Joy (4), Stress (2) and No-Idea (1)
5	22,24, 24,25	3 Males, 1 Female	19.7, 19.6, 21, 22.3	Normal (3), Sad (3), Dislike (4), Joy (4), Stress (3) and No-Idea (2)
6	24,25, 25,27	4 Males	25, 19.2, 21.7, 20.9	Normal (2), Sad (4), Dislike (4), Joy (4), Stress (3) and No-Idea (0)

We analysed the dataset Type 1 and the dataset Type 2 in three different ways. As a result of pre-tests on choosing an appropriate machine learning algorithm we decided to use J48, the open source Java implementation of the C4.5 decision tree algorithm [44] in the WEKA tool for the analysis of the different datasets. To reduce variability, multiple rounds of cross-validation were performed using different partitions and the validation results were combined (e.g. averaged) over the rounds to estimate a final predictive model. This was done instead of using the conventional validation (e.g. partitioning the data set into two sets of 70% for training and 30% for test). We used 10-fold cross-validation to partition the data into separate training and test sets without losing significant modelling or testing capability.

The J48 machine learning algorithm we used with the dataset of each participant ('Individual') and applied it to the dataset of each of the six above mentioned groups as shown in tab. 9.3 ('Group'). To manage the limitation of the implementation in processing the complete dataset, we chose portions of data randomly pertaining to each emotional state ('Portioned').

Out of 50.000 instances of emotional states included in our datasets, we got two types of data i.e. 'Six-class" and 'Two-class"; each type was analysed on 'Individual', 'Group' and 'Portioned' basis. To identify the appropriate choice and combination of sensors we used the different datasets starting with the data of all four sensors of blood volume pulse (BVP), galvanic skin response (GSR), electromyography (EMG) and skin temperature (ST). In further steps we took just one, two or three sensors out to see how the recognition rate changes and which seems to be the appropriate choice.

Table 4 Correlation between Type 1 categories of IAPS images and result with BVP, GSR, EMG and ST sensors

4 sensors	accuracy	min	max	SD	sad	dislike	joy	Stress	Normal	no idea
individual	99.1	98.4	99.5	0.25	99.0	99.1	99.0	99.2	99.2	98.9
group	98.7	98.3	99.0	0.26	98.5	98.8	98.6	98.8	98.7	98.6
partitioned	98.5				98.2	98.8	98.4	98.3	98.6	98.0

The results for the four sensors and the six emotional states (Type 1) are shown in table 4. With a probability of over 98 % (standard deviation (SD) of ~0.25%) we can recognize based on the data collected by the four sensors any of the emotional states (Type 1) indicated by the IAPS images with more than 98% probability.

When looking only for a 'Positive' or 'Negative' emotional state (Type 2) we obtained the results as shown in table 5. In this case, the recognition rate is even ~ 99%, which means by analysing the data of the 4 sensors one can reliably detect either a positive or negative emotional state, independent whether we analyse the individual or group dataset or even an arbitrary dataset just taken out of a sampling.

Table 5 Correlation between Type 2 categories of IAPS images and result with BVP, GSR, EMG and ST sensors

4 sensors	accuracy	min	max	SD	positive	negative
individual	99.4	97.7	99.7	0.45	98.9	99.6
group	99.3	99.1	99.5	0.14	98.7	95.6
partitioned	99.3				98.6	99.7

Table 6 Correlation between Type 2 categories of IAPS images and result with three sensors out of BVP, GSR, EMG and ST

3 sensors		accuracy	min	max	SD	positive	negative
BVP	individual	99.3	98.7	99.7	0.27	98.7	99.6
GSR	group	99.2	98.9	99.5	0.20	98.5	99.6
ST	partitioned	99.3				98.3	99.7
BVP	individual	99.1	97.8	99.7	0.39	98.2	99.5
EMG	group	98.9	98.6	99.2	0.21	97.8	99.4
ST	partitioned	97.5				91.1	97.9
BVP	individual	98.1	94.0	99.7	1.41	96.4	99.0
EMG	group	97.6	96.4	98.2	0.63	95.2	98.7
GSR	partitioned	97.8				95.1	98.9
EMG	individual	99.4	98.6	99.7	0.32	98.7	99.6
GSR	group	99.2	98.9	99.4	0.21	98.4	99.5
ST	partitioned	99.3				98.5	99.7

To see if the recognition rate remains sufficient when eliminating the data of one sensor out of the data set a further sequence of analysis was done. The result is shown in table 6. As long as the data of the skin temperature sensor (ST) is included, the recognition rate remains at ~99%.

To see if that also holds true when we look for the six Type 1 emotional states we did a further analysis as shown in table 7. Here it is also obvious that the data of the skin temperature sensor (ST) improves the recognition rate to ~ 98%. This led to the consideration to reduce the number of sensors from which the collected data was included even further.

Table 7 Correlation between Type 1 categories of IAPS images and result with three sensors out of BVP, GSR, EMG and ST

3 sensors		accu-racy	min	max	SD	sad	dis-like	joy	str-ess	Nor-mal	no idea
BVP	individual	98.8	97.8	99.6	0.53	98.6	99.0	98.9	98.9	98.8	98.9
GSR	group	98.6	98.2	98.9	0.31	98.3	98.7	98.5	98.6	98.5	98.7
ST	partitioned	98.6				98.3	99.0	98.4	98.6	98.7	98.4
BVP	individual	98.5	96.5	99.2	0.64	98.2	98.7	98.3	98.6	98.7	98.8
EMG	group	98.0	97.6	98.5	0.34	97.7	98.3	97.9	98.1	98.2	98.3
ST	Partition.	98.1				97.8	98.4	97.9	97.9	98.4	97.8
BVP	individual	96.6	91.4	99.3	2.82	96.4	96.9	95.9	96.4	97.4	97.3
EMG	group	96.0	94.5	96.7	0.79	95.8	96.2	95.3	95.8	96.9	96.7
GSR	Partition.	96.5				96.8	98.8	96.0	96.5	96.7	96.8
EMG	individual	98.7	96.9	99.6	0.78	98.4	98.9	98.6	98.8	99.0	98.5
GSR	group	98.4	97.7	98.9	0.42	98.0	98.6	98.4	98.4	98.6	98.2
ST	partitioned	98.6				98.5	98.9	98.4	98.4	99.0	97.9

In table 8, the results of this analysis show that one can achieve out of the sensor signals by the combination of the skin temperature (ST) and the BVP or EMG with a recognition rate of ~95% 'positive' and with even ~98% reliability the 'negative' emotional state of the person wearing those sensors. EMG delivers slightly better results than BMP.

Table 8 Correlation between Type 2 categories of IAPS images and result with two sensors out of BVP, GSR, EMG and ST

2 sensors		accuracy	min	max	SD	positive	negative
BVP GSR	individual	99.9	87.8	97.8	2.39	89.5	97.4
	group	94.7	92.7	96.3	1.17	89.2	97.2
	partitioned	95.5				90.8	97.6
BVP ST	individual	97.5	92.9	99.5	1.69	94.9	98.7
	group	97.5	96.9	98.1	0.48	95.0	98.7
	partitioned	98.2				96.4	98.9
BVP EMG	individual	95.8	88.0	99.6	3.44	91.3	97.7
	group	95.5	93.3	98.0	1.74	91.0	97.6
	partitioned	96.0				91.9	97.9
EMG GSR	individual	95.4	87.6	98.7	2.57	91.1	97.4
	group	95.0	93.1	96.4	1.31	90.1	97.3
	partitioned	96.4				92.4	98.1
EMG ST	individual	98.2	93.8	99.3	1.12	96.2	99.1
	group	97.9	99.8	98.4	0.55	95.6	98.9
	partitioned	98.3				96.5	99.2
GSR ST	individual	95.7	89.8	98.3	2.66	91.1	97.9
	group	94.9	92.8	97.3	1.55	86.6	97.8
	partitioned	96.1				90.3	98.7

Table 9 Correlation between Type 1 categories of IAPS images and result with two sensors out of BVP, GSR, EMG and ST

2 sensors		accu-racy	min	max	SD	sad	dis-like	joy	Str-ess	Nor-mal	no idea
BVP GSR	individual	91.1	79.0	96.8	4.98	90.7	92.0	90.0	90.2	93.4	91.4
	group	90.2	81.8	93.3	2.20	89.8	91.4	89.1	89.3	92.5	90.3
	partitioned	92.6				92.9	93.9	90.8	91.8	93.6	93.0
BVP ST	individual	96.2	88.8	99.1	2.44	95.6	97.1	95.6	96.2	94.9	96.8
	group	96.0	94.8	96.4	0.77	95.0	96.6	95.2	95.5	94.5	96.2
	partitioned	96.4				95.2	97.1	96.6	96.8	95.1	96.8
BVP EMG	individual	93.4	76.6	99.4	6.07	92.6	93.8	91.2	93.1	96.0	98.9
	group	92.8	89.2	96.6	2.89	92.4	93.2	91.1	92.3	95.6	96.4
	partitioned	93.9				93.2	94.2	92.2	94.1	97.8	96.8
EMG GSR	individual	91.6	80.6	91.6	4.27	99.0	99.1	99.0	99.2	99.2	98.9
	group	90.8	87.9	93.0	1.93	88.8	92.6	89.7	91.0	92.2	89.9
	partitioned	93.6				93.8	94.1	92.2	93.2	95.5	92.7
EMG ST	individual	96.7	89.7	99.4	2.14	95.9	97.2	96.4	97.0	97.0	95.6
	group	95.9	94.3	96.6	0.83	95.2	96.5	95.7	95.9	96.1	95.0
	partitioned	97.0				97.0	98.0	95.8	96.3	97.9	96.8
GSR ST	individual	91.6	79.2	91.6	4.93	91.3	93.0	90.8	92.0	91.4	87.2
	group	89.9	87.7	94.5	2.58	89.1	92.0	89.4	90.0	87.9	86.2
	partitioned	91.3				89.4	95.8	88.5	91.0	90.6	82.5

This also holds for the six Type 1 emotional states as shown in table 9. The recognition rate of the different emotional states just using the data of two sensors (BVP or EMG together with ST) is however not as good as with the Type 2 states, but still above 95% recognition rate for the most important states of 'dislike' and 'stress'. Again EMG with ST deliver slightly better results than BV with ST.

Table 10 Correlation between Type 2 categories of IAPS images and result with one sensor out of BVP, GSR, EMG and ST

1 sensor		accuracy	min	max	SD	positive	negative
BVP	individual	86,8	69,3	94,1	6,25	69,9	94,2
	group	85,3	83,3	88,2	1,74	67,4	93,5
	partitioned	86,5				70,6	93,5
EMG	individual	86,6	74,7	95,3	5,69	75,3	94,4
	group	87,1	85,2	89,8	2,03	72,3	93,8
	partitioned	89,4				78,3	94,3
GSR	individual	80,3	71,4	87,3	4,07	54,0	92,7
	group	78,6	75,9	82,4	2,16	46,0	93,7
	partitioned	79,7				48,4	93,6
ST	individual	89,5	76,8	96,8	5,89	75,2	96,1
	group	86,6	82,9	91,9	3,16	66,7	95,8
	partitioned	88,6				69,5	97,5

Table 11 Correlation between Type 1 categories of IAPS images and result with one sensor out of BVP, GSR, EMG and ST

1 sensor		accu-racy	min	max	SD	sad	dis-like	joy	Str-ess	Nor-mal	no idea
BVP	individual	74,8	54,1	87,8	11,21	73,1	78,3	67,7	73,3	81,0	78,0
	group	71,3	65,4	77,2	4,69	69,8	75,6	65,8	68,8	76,7	73,3
	partitioned	73,9				71,3	79,5	70,7	66,7	80,3	75,6
EMG	individual	78,9	55,5	91,0	10,37	75,3	82,0	73,2	82,1	80,8	82,2
	group	75,8	71,1	80,4	3,60	72,4	80,0	70,4	76,6	77,8	78,7
	partitioned	80,3				79,3	82,9	78,4	76,1	84,0	83,3
GSR	individual	60,5	43,3	83,6	8,82	56,9	69,2	48,8	66,2	64,3	44,9
	group	55,9	50,0	66,7	3,99	49,5	73,1	39,6	58,0	51,2	43,2
	partitioned	54,4				45,4	82,2	41,7	43,1	46,5	17,8
ST	individual	79,2	61,6	91,9	9,81	77,4	82,8	73,7	82,9	78,6	72,8
	group	72,4	68,2	79,8	4,30	65,8	86,0	69,0	67,2	64,8	63,1
	partitioned	74,7				68,8	91,3	66,0	73,0	68,7	42,3

Finally, we checked if, even with only one sensor, reliable information can be found (see table 10 and 11). However, the results are not very promising. Although the recognition rate of negative emotional states are all with just one sensor above 90% and with the skin temperature sensor (ST) even ~96% but those of positive emotional states are all below 80% (see table 10). If one looks at the six Type 1 categories the recognition rates are not satisfactory for any judgment at all.

5   Conclusion and Future Work

In the experiments, the emotional states of 24 participants were stimulated using IAPS images [32]. We changed the emotional state with these images and the participants had to decide between 'sad', 'dislike', 'joy', 'stress', 'normal', and 'no idea' to express their 'positive' or 'negative' emotional state while being monitored using four different physiological sensors (BVP, EMG, GSR and ST).

Table 12 Summary of Type 1 categories of IAPS images and result with different sensor configurations of BVP, GSR, EMG and ST

Sensors	Correctly classified instances (Individual) Average	Correctly classified instances (Groups) Average	Chunks from all participants
EMG, BVP, GSR, ST	99.1	98.7	98.5
BVP, ST, GSR	98.8	98.6	98.6
EMG, BVP, ST	98.5	98.0	98.1
EMG, BVP, GSR	96.6	96.0	96.5
EMG, GSR, ST	98.7	98.4	98.6
BVP, GSR	91.1	90.2	92.6
BVP, ST	96.2	96.0	96.4
EMG, BVP	93.4	92.8	93.9
EMG, GSR	91.6	90.8	93.6
EMG, ST	96.7	95.9	97.0
GSR, ST	91.6	89.9	91.3
BVP	74.8	71.3	73.9
EMG	78.9	75.8	80.3
GSR	60.5	55.9	54.4
ST	79.2	72.4	74.7

We then analysed the data in the following way: We took data of each participant and applied J48 classifier and an average of 'individual' data. The data of six participants were classified in the same way with the result of 'group' data. Small portions of data randomly chosen from each participant were also analysed with the J48 classifier as 'partitioned' data. We categorized the data in six emotional states i.e. sad, dislike, joy, stress, normal and no-idea (Type 1) and also the two emotional states i.e. positive and negative (Type 2). Using different combinations of physiological sensors in order to see the importance of each of them helped us to find the right choice of sensors.

Table 12 represents the results for the identification of all six emotional states (Type 1) by using the IAPS images for stimulation and the sensor signals for detection. According to this, the accuracy decreased when we used less physiological sensors. The best results we achieved when taking all four sensors into account. The results were still above 98 % when we used ST together with two of the other sensors. When using only two sensors ST gave best results together with either BVP or EMG. With only one sensor one cannot achieve reliable detection of the emotional state.

Table 13 Summary of Type 2 categories of IAPS images and result with different sensor configurations of BVP, GSR, EMG and ST

Sensors	Correctly classified instances (Individual); Average	Correctly classified Instances (Groups); Average	Chunks from all participants
EMG, BVP, GSR, ST	99.4	99.3	99.3
BVP, ST, GSR	99.3	99.2	99.3
EMG, BVP, ST	99.1	98.9	99.0
EMG, BVP, GSR	98.1	97.6	97.8
EMG, GSR, ST	99.4	99.2	99.3
BVP, GSR	94.9	94.7	95.5
BVP, ST	97.5	97.5	98.1
EMG, BVP	95.8	95.5	96.0
EMG, GSR	95.4	95.0	96.4
EMG, ST	98.2	97.9	98.3
GSR, ST	95.7	94.9	96.1
BVP	86.8	85.3	86.5
EMG	88.6	87.1	89.4
GSR	80.3	78.6	79.7
ST	89.5	86.6	88.6

When only looking for either 'positive' or 'negative' emotional states (Type 2) with the definition that 'positive' means 'joy' and 'normal' and 'negative' means 'stress', dislike' and 'sad' one achieves, with the same approach for the analysis of the merged data, similar but slightly better results as shown in table 13. The best results (>99%) we achieved when taking all four sensors into account. The results were still above 99 % when we used ST together with two of the other sensors. When using only two sensors, ST gave best results (>97%) together with either BVP or EMG. With only one sensor one cannot achieve reliable detection of the emotional state at all also here.

The approach presented here with the test set-up as shown in figure 1 was able to recognize the aforementioned emotional states by using physiological devices and the J48 implementation of the C4.5 decision tree classifier with high accuracy.

With the right two physiological devices measuring the skin temperature (ST) and either the blood volume pulse (BVP) or electromyograph (EMG) one can recognize emotional states like sad, dislike, joy, stress, normal, no-idea, positive and negative. The prototype we used is only a 'proof of concept' and the results show that the approach is appropriate. We had some variance in the body mass index (BMI) and age group. However, this needs further investigations. The physiological sensors needed a proper fixture on the participants' skin in order to gain reliable signals. Here further development is needed to get wireless sensing for a better form factor. It will furthermore be helpful to research facial expressions for recognizing emotional states.

6   References

[1]   Cooking Hacks: e-Health Sensor Platform V2.0 for Arduino and Raspberry Pi [Biometric / Medical Applications]. http://www.cooking-hacks.com/documentation/tutorials/ehealth-biometric-sensor-platform-arduino-raspberry-pi-medical#step4_9

[2]   Professor Rosalind W. Picard.
       http://web.media.mit.edu/~picard/index.php

[3]   Affective Computing: Publications.
       http://affect.media.mit.edu/publications.php

[4]   Richmond Hypnosis Center.
       http://richmondhypnosiscenter.com/2013/04/ 12/sample-post-two/

[5]   American Psychological Association: The Impact of Stress.
       http://www.apa. org/news/press/releases/stress/2011/impact.aspx

[6]   WebMD: Stress Management Health Center.
        http://www.webmd.com/balance/stress-management/effects-of-
        stress-on-your-body

[7]   Krifka, S., Spagnuolo, G., Schmalz, G., Schweikl, H. A review of adaptive
        mechanisms in cell responses towards oxidative stress caused by den-
        tal resin monomers. Biomaterials. 2013; 34: pp. 4555–4563.

[8]   Healthline: The Effects of stress on the Body.
        http://www.healthline.com/health/stress/effects-on-body

[9]   Miamiherald: Chronic stress is linked to the six leading causes of death.
        http://www.miamiherald.com/living/article1961770.html

[10]   Online Stress Reliever Games.
        http://stress.lovetoknow.com/Online_ Stress_Reliever_Games

[11]   Ali Mahmood Khan. (2011). Personal state and emotion monitoring by
        wearable computing and machine learning. BCS-HCI 2012, Newcastle,
        UK

[12]   WebMD: Stress Management Health Center.
        http://www.webmd.com/balance/stress-management/features/how-
        anger-hurts-your-heart

[13]   BetterHealth: Anger - how it affects people.
        http://www.betterhealth.vic.gov.au/bhcv2/bhcarticles.nsf/pages/Ange
        r_how_it_affects_people

[14]   Ekman, Paul (1999), "Basic Emotions", in Dalgleish, T; Power, M, Hand-
        book of Cognition and Emotion (PDF), Sussex, UK: John Wiley & Sons.

[15]   Health and Wellness: Are Happy People Healthier? New Reasons to
        Stay Positive.
        http://www.oprah.com/health/How-Your-Emotions-Affect-Your-
        Health-and-Immune-System

[16]   Darwin, C. (1872). The expression of the emotions in man and animals.
        John Murray, London

[17]   Kok, B.E., Coffey, K.A., Cohn, M.A., Catalino, L.I., Vacharkulksemsuk, T.,
        Algoe, S., Brantley, M. & Fredrickson, B. L. (2013). How positive emo-
        tions build physical health: Perceived positive social connections ac-
        count for the upward spiral between positive emotions and vagal tone.
        Psychological Science, 24(7), pp. 1123-1132.

[18]   Atkinson, William Walker (1908). Thought Vibration or the Law of At-
        traction in the Thought World.

[19]   Rush, A. J., Beck, A. T., Kovacs, M. & Hollon, S. D. Comparative efficacy
        of cognitive therapy and pharmacotherapy in the treatment of de-
        pressed outpatients. Cognit. Ther. Res. 1, pp. 17-38.

[20] Remarks on Emotion Recognition from Bio-Potential Signals: research paper (2004)

[21] Emotion Recognition from Electrocardiogram Signals using Hilbert Huang Transform (2012)

[22] Emotion Pattern Recognition Using Physiological Signals (2014)

[23] Walter Sendlmeier Felix Burkhardt, Verification of Acoustical Correlates of Emotional Speech using Formant-Synthesis, Technical University of Berlin, Germany.

[24] RECOGNIZING EMOTION IN SPEECH; Frank Dellaert, Thomas Polzin and Alex Waibel

[25] RECOGNIZING EMOTION IN SPEECH USING NEURAL NETWORKS; Keshi Dai, Harriet J. Fell and Joel MacAuslan

[26] Recognizing Emotion from Facial Expressions: Psychological and Neurological Mechanisms Ralph Adolphs, University of Iowa College of Medicine.

[27] Analysis of emotion recognition using facial expressions, speech and multimodal information. Proceedings of the 6th International Conference on Multimodal Interfaces, ICMI 2004, State College, PA, USA, October 13-15, 2004

[28] Recognizing Emotions from Facial Expressions Using Neural Network; Isidoros Perikos, Epaminondas Ziakopoulos, Ioannis Hatzilygeroudis.

[29] Emotions States Recognition Based on Physiological Parameters by Employing of Fuzzy-Adaptive Resonance Theory; Mahdis Monajati, Seyed Hamidreza Abbasi, Fereidoon Shabaninia, Sina Shamekhi.

[30] Classification of emotional states from electrocardiogram signals: a non-linear approach based on hurst; Jerritta Selvaraj, Murugappan Murugappan, Khairunizam Wan and Sazali Yaacob

[31] Jennifer Healey and Rosalind W. Picard (2002), Eight-emotion Sentics Data, MIT Affective Computing Group, http://affect.media.mit.edu."

[32] THE CENTER FOR THE STUDY OF EMOTION AND ATTENTION: http://csea.phhp.ufl.edu/Media.html#topmedia

[33] Cuthbert BN, Schupp HT, Bradley MM, Birbaumer N, Lang PJ.; Brain potentials in affective picture processing: covariation with autonomic arousal and affective report. Biol Psychol. 2000 Mar;52(2): 95-111.

[34] Keil A, Bradley MM, Hauk O, Rockstroh B, Elbert T, Lang PJ.; Large-scale neural correlates of affective picture processing.;Psychophysiology. 2002 Sep; 39(5): pp. 641-649.

[35] Lang PJ, Bradley MM, Fitzsimmons JR, Cuthbert BN, Scott JD, Moulder B, Nangia V.; Emotional arousal and activation of the visual cortex: an fMRI analysis. Psychophysiology. 1998 Mar; 35(2): pp. 199-210.

[36]   Bradley MM1, Sabatinelli D, Lang PJ, Fitzsimmons JR, King W, Desai P.;
        Activation of the visual cortex in motivated attention.; Behav Neurosci.
        2003 Apr; 117(2): pp. 369-380.

[37]   Sabatinelli D, Bradley MM, Fitzsimmons JR, Lang PJ.; Parallel amygdala
        and inferotemporal activation reflect emotional intensity and fear rel-
        evance.; Neuroimage. 2005 Feb 15; 24(4): pp. 1265-70. Epub 2005 Jan
        7.

[38]   Sabatinelli D, Lang PJ, Keil A, Bradley MM.; Emotional perception: cor-
        relation of functional MRI and event-related potentials.; Cereb Cortex.
        2007 May; 17(5): pp. 1085-1091. Epub 2006 Jun 12.

[39]   Bradley MM, Codispoti M, Lang PJ.; A multi-process account of startle
        modulation during affective perception.; Psychophysiology. 2006
        Sep;43(5): pp. 486-497.

[40]   Maker Shed: Pulse Sensor AMPED for Arduino.
        http://www.makershed.com/products/pulse-sensor-amped-for-
        arduino

[41]   Raspberry Pi. https://www.raspberrypi.org/

[42]   Revathi priya Muthusamy, "Emotion Recognition from Physiological
        signals using Bio-sensors",
        https://diuf.unifr.ch/main/diva/sites/diuf.unifr.
        ch.main.diva/files/T4.pdf - Submitted for Research Seminar on Emo-
        tion Recognition on 15.02.2012.

[43]   Russel and Bullock, Multidimensional scaling of emotional facial ex-
        pressions, Journal of Personality and Social Psychology (1985), pp.
        1290-1298

[44]   Quinlan, J. R. C4.5: Programs for Machine Learning. Morgan Kaufmann
        Publishers, 1993.

X   List of Authors

Dr. Antonio Ascolese holds a PhD in 'Subjective well-being, health and cross cultural communication' from the University of Milano-Bicocca, graduating in Developmental and Communication Psychology (Master's degree) in 2006 at the Catholic University of Milan. He has been working as Senior Researcher at CESCOM (Centre for Research in Communication Science) at the University of Milano-Bicocca, and as a qualitative researcher in market research. His research interests include positive psychology, psychology of nonverbal communication, psychology of emotions and learning processes in simulated environments. He is currently working at imaginary srl, as a Serious Games project manager, on European research based projects. He is also professor of the course 'Media Psychology', at the Sigmund Freud University, in Milan. He has been a Chartered Psychologists since 2007 and cognitive-behavioural psychotherapist since 2016.

Msc Rita Bertoni works for Fondazione Don Gnocchi since 2009 as a researcher and physical therapist. During these years she has collaborated at different research projects involving the use of end-effector robot and myoelectrically controlled functional electrical stimulation for the rehabilitation of the upper limb. She obtained her Msc in sports sciences at Università Cattolica del Sacro Cuore in 2005 and graduated as a Physical Therapist at Università degli Studi di Milano in 2008.

Prof. Dr. Silvana Dellepiane graduated in Electronic Engineering in 1986 with honours. In 1990 she received the PhD degree in Electronic Engineering and Computer Science and in 1992 she became a Researcher (Assistant Professor) in the Department of Biophysical and Electronic Engineering (DIBE), Università degli Studi di Genova. She is now there an Associated Professor in the Department of Naval Electrical Electronical and Telecommunication Engineering teaching Signal Theory, Statistical Methods, and Pattern Recognition. At present she is professor of Electrical Communications, Image Processing and Recognition, in the courses of Telecommunications Engineering, Bioengineering, and Cultural Heritage. Her main research interests include the use of context and fuzzy systems for multi-dimensional and multi-temporal data processing, segmentation, supervised methods for the processing of remote sensing SAR images, and non-linear adaptive processing of digital signals. Her major application domains are telemedicine and remote sensing. Since 2011 she is involved as coordinator for cooperation projects towards Africa and developing

© Springer Fachmedien Wiesbaden GmbH, part of Springer Nature 2018
M. Lawo und P. Knackfuß (Hrsg.), *Clinical Rehabilitation Experience Utilizing Serious Games*, Advanced Studies Mobile Research Center Bremen, https://doi.org/10.1007/978-3-658-21957-4

countries, as well as coordinator of the Polytechnic Engineering School for the Interschool Cooperation course, established by Università di Genova.

Marten Ellßel, MSc. graduated from the University of Bremen in Systems Engineering (Bsc) and 2014 in Computer Science (Msc). He was involved in several projects of interaction design and mobile computing. He started working as a Researcher and Android Developer for Neusta mobile solutions and is now Android Lead Developer at Open Reply in Bremen.

Dr. Roberta Ferretti graduated 2012 (Bsc) and 2014 (Msc) in Biomedical Engineering at University of Genoa. Since 2013 she is involved in research activities concerning Image Processing and Data Fusion algorithms at the Department of Naval Electrical Electronical and Telecommunication Engineering there. In 2017 she obtained her Ph.D. in Science and Technology for Electronics and Telecommunication Engineering: her research topics are techniques of nonlinear processing, segmentation, data fusion and analysis of the quality of digital images. The application domain is biomedical field.

Dov (Dubi) Faust holds an Msc in Organo-Physical Chemistry of Tel-Aviv University and is a strategic business consultant. He has a wealthy senior management experience of over 30 years in people, technology and operation arenas in large multidisciplinary, global hi-tech companies. As former HR Director at Micron Israel and Operation manager at Intel Israel he is an expert in organizational effectiveness aspects like strategic planning and strategy deployment, value-stream Improvement of core organizational processes, organizational culture and knowledge management. Lately he was In charge of EU related programs consolidation and coordination as a member of Edna Pasher & Ass. Team.

Elisa Ferrara obtained the Law degree (BA) in 2009 at University of Genova and the MA on International Sciences in 2012 at University of Turin. Since 2013 she is a Fellow Researcher in design and assessment of sociological technology services in the field of health and rehabilitation at the Department of Naval Electrical Electronical and Telecommunication Engineering. Her research topics are concentrated on international regulation and protocol for health data management, interaction and assessment of sociable technologies in health care and rehabilitation.

Dr. Hendrik Iben worked from 2007 till 2015 at the TZI (www.tzi.de) of the University of Bremen as a research assistant. He was involved in numerous projects of wearable computing and eHealth. He is a 2007 graduate of computer science of Universitaet Bremen and received his PhD in 2015 at the same university. Since 2015 he is working at Ubimax GmbH (www.ubimax.com) as a lead software engineer.

Dr. Johanna Jonsdottir is at Fondazione Don Gnocchi since 2004 as a senior researcher. During these years she has led or collaborated in numerous research projects involving persons with neurological disorders, including rehabilitation of gait and reaching, validation of outcome measures, and development and validation of the use of virtual reality and robotics in rehabilitation. She obtained her MSc and DSc in Motor control and Applied Kinesiology from Boston University, Boston Massachusetts in 1999. She is Docent in Physical Therapy at University of Milan.

Dr. Ali Mehmood Khan is a 2010 graduate of Computer Science at the University of Bremen (MSc) and worked as a research associate at TZI of Bremen University from 2012 to 2016 in a couple of public funded projects in the domain of eHealth, Wearable Computing and AAL. He received his PhD from Bremen Universität in 2017. Besides that he has worked and now works as a Software developer in German software companies for few years.

Dr. Wolfhard Klein is neuropsychologist at the Neurologisches Therapiezentrum Gmundnerberg, a neurologic rehabilitation centre, since 2010 as coordinator of special projects. He is a 1993 graduate of Psychology of Universität Salzburg and received his PhD from Universität Salzburg in 2003. From 1994 to 1997 and from 2000 to 2003 he was involved in scientific projects in the field of human motor control at the Institute for Labour Physiology at the Universität Dortmund.

Dr. Peter Knackfuß holds German diploma degrees in mechanical and electrical engineering. After his studies he joined Gesellschaft für Reaktorsicherheit in Munich working as a software engineer. 1987 he changed to the 'Bremen Institute for Industrial Engineering and Work Science – BIBA' starting as a research engineer and working as a department leader later on. He graduated as a Dr. Engineer at the University of Bremen in 1992. In 1997 he founded his own company dealing with network technology and the support of Industry and Public Bodies while performing research and development projects.

Prof. Dr. Michael Lawo was from 2004 till his retirement at the end of 2016 at TZI (www.tzi.de) of the University of Bremen as professor for applied computer science. He was leading and involved in numerous projects of human robot interaction, logistics, wearable computing, eHealth, AAL and artificial intelligence. He is a 1975 graduate of structural engineering of Ruhr Universität Bochum, received his PhD from Universität Essen in 1981 and became professor in structural optimization there in 1992. In 2000 he was appointed as professor of honour of the Harbin/China College of Administration & Management. His research focus is on human machine interaction. Beside the scientific career he looks back on more than 30 years of management experience in industry, till 2016 as general manager of neusta mobile solutions. He now is consultant of two research institutions.

Prof. Robert K. Logan, Professor Emeritus Physics, Fellow of St. Michael's College, U. of Toronto, and Chief Scientist at OCAD U. (Ontario College of Art and Design) in the Strategic Innovation Lab, has a variety of experiences as an academic involved in research in media ecology, complexity theory, information theory, systems biology, environmental science, linguistics, and industrial design. He has published with and collaborated with Marshall McLuhan and continues his McLuhan studies research. He is also an author or editor of 16 books and over 100 articles in refereed journals and a Senior Fellow at the Origins Institute, McMaster U., a Senior Fellow at the Institute of Biocomplexity and Informatics, U. of Calgary, a fellow of the Bertalanffy Center for the Study of Systems Science and an emeritus faculty of the School of the Environment (U of Toronto/Canada). In June 2011 he was presented with the Walter J. Ong Award for Career Achievement in Scholarship by the Media Ecology Association.

Dr. Sonia Nardotto graduated in Biomedical Engineering at University of Genoa in 2010 (BSc) and 2012 (MSc) with a thesis on 'An automatic method for segmentation of multiparametric medical volumes". Since then she is in research activities concerning Image Processing and Data Fusion algorithms at the Department of Naval Electrical Electronical and Telecommunication Engineering. In 2017 she obtained a Ph.D. in Information and Knowledge Science and Technology. Her research topics are techniques of non-linear processing, segmentation, data fusion and analysis of the quality of digital images.

Lucia Pannese is CEO and owner of imaginary in Milan. She graduated in applied mathematics in 1993. She has more than 20 years managing experience in international innovation & research projects dealing with Serious Games, gamification, virtual reality and enabling technologies for eHealth, training and behavioural change.

Dr. Edna Pasher earned 1981 her PhD at New York University in Communication Arts and Sciences and has served as faculty member at Adelphi University, the City University of New York, the Hebrew University in Jerusalem and the Tel-Aviv University. Edna has been a pioneer and leader of the innovation and knowledge management movement in Israel and an active member of the international community of the KM pioneers. She has over 30 years of experience in regional and international IS projects using a variety of evaluation methodologies, modelling techniques, and quantitative analysis. She is the founder and chair of Edna Pasher PhD & Associates (1978) and of Status the Israeli management magazine (1991) and of the Israeli Smart Cities Institute (2015).

Dr. Serena Ponte graduated in Biomedical Engineering at University of Genoa in 2012 (BSc) and 2014 (MSc) with a thesis on 'Biofeedback methods for monitoring patients during rehabilitation at home'. Since 2014 she is a Ph.D. student in Science and Technology for Electronic and Telecommunications Engineering at the Department of Naval Electrical Electronical and Telecommunication Engineering. Her research topics focus on biofeedback methods for the analysis of biomedical signals and images for monitoring and follow up. In particular, she studies the heartbeat during a rehabilitation program for patients with cognitive problems that lead to motor disabilities of the upper limb.

Hadas Raz was management consultant and researcher at Edna Pasher PhD & Associates from 2010 until 2016. She has a B.A. in Social science from Ben-Gurion University, Israel (2007) and a M.A. in Organizational and occupational psychology of Haifa University (2011). She was involved in projects on qualitative organizational diagnosis, organizational change, development and learning, strategic planning and training. She has experiences as research assistant in social sciences within projects funded by the European Commission and the Israeli Democracy Institute. Since 2017 she is at the Center for Educational Technology in Tel Aviv/Israel.

David Wortley is the former Founding Director of the Serious Games Institute (SGI) at Coventry University which he set up between 2007 and 2011. He is currently President of the European Chapter of the International Society of Digital Medicine and an acknowledged expert practitioner of wearable technologies and gamification for personal health management. He has worked with Imaginary as a consultant since 2011.

Printed in the United States
By Bookmasters